W0036076

OUR TRAUMATIZED* PLANET

* **trauma** ---- an injury, physical, or psychological

A Stark Perspective on the Earth's Environmental Crises

deforestation,
disease,
famine,
genocide,
greed,
lies,
mass extinctions,
overexploitation,
plastic overload,
poisoned water,
polluted air,
population issues,
poverty,
war,
wildfires,

oh, and lest we forget … the climate crisis

What We Can Learn and Apply from Contemporary Traditional Peoples,
Ancient Societies, and our own Successes and Failures

Mark Q. Sutton and E. N. Anderson

OUR TRAUMATIZED* PLANET

* trauma ---- an injury, physical, or psychological

A Stark Perspective on the Earth's Environmental Crises

Mark Q. Sutton and E. N. Anderson

Routledge
Taylor & Francis Group

LONDON AND NEW YORK

Designed cover image: Boulder Glacier, Washington State, 1932 and 1988

First published 2025
by Routledge
4 Park Square, Milton Park, Abingdon, Oxon OX14 4RN

and by Routledge
605 Third Avenue, New York, NY 10158

Routledge is an imprint of the Taylor & Francis Group, an informa business

© 2025 Mark Q. Sutton and E. N. Anderson

The right of Mark Q. Sutton and E. N. Anderson to be identified as authors of this work has been asserted in accordance with sections 77 and 78 of the Copyright, Designs and Patents Act 1988.

All rights reserved. No part of this book may be reprinted or reproduced or utilised in any form or by any electronic, mechanical, or other means, now known or hereafter invented, including photocopying and recording, or in any information storage or retrieval system, without permission in writing from the publishers.

Trademark notice: Product or corporate names may be trademarks or registered trademarks, and are used only for identification and explanation without intent to infringe.

British Library Cataloguing-in-Publication Data
A catalogue record for this book is available from the British Library

ISBN: 9781032908922 (hbk)
ISBN: 9781032898995 (pbk)
ISBN: 9781003560326 (ebk)

DOI: 10.4324/9781003560326

Typeset in Times New Roman
by codeMantra

CONTENTS

ABOUT THE BOOK

Our Traumatized Planet explores the state of the environment and some of the major issues faced today and asks what we can learn and apply from contemporary traditional peoples, ancient societies, and our own successes and failures.

Providing straightforward information on some of the serious environmental issues we face so that non-scientists can understand them, this book explores what is at stake so that we can choose to make a difference. Combining the latest data from environmental, anthropological, and archaeological science allows for fresh perspectives and an empirical approach to describing these problems that eliminates hopeful denial, speculation, wishful thinking, and downright lies. Using archaeological data, the authors provide examples of success and failures in the past that could be used to make decisions about the future. They also highlight examples of how traditional peoples, past and present, have dealt with these same issues. Seeing the current crisis through the eyes of two experienced anthropologists broadens our understanding and allows us to set contemporary issues in the context of the past and traditional knowledge. However, this is not a book of easy solutions from the past to solve our future; rather, it is an impassioned plea to people today to read and understand what state the planet is in and encourage them to find the will to change.

This book is for students of archaeology, anthropology, and environmental science and all those wanting to, in a clear and readable way, understand the fate of our planet.

ABOUT THE AUTHORS

Mark Q. Sutton is Professor of Anthropology, Emeritus, at California State University, Bakersfield. He received his Ph.D. in Anthropology from the University of California, Riverside, in 1987. Dr. Sutton specializes in prehistory, hunter-gatherer adaptations to arid environments, insects as food and in technology, prehistoric diet and technology, and ecology. Dr. Sutton has published more than 250 books, monographs, articles, and reviews on archaeology and anthropology, including the textbooks *Laboratory Methods in Archaeology* (1996, 7th ed., 2019), *Introduction to Native North America* (2000, 7th ed., 2024), *Archaeology: The Science of the Human Past* (2003, 7th ed., 2024), *Introduction to Cultural Ecology* (with E. N. Anderson, 2004, 3rd ed., 2014), *Paleonutrition* (2010), *A Prehistory of North America* (2011), *Bioarchaeology* (2021), and *A Concise Introduction to Cultural Anthropology* (2022). He lives in San Diego with his wife, Melinda, and their dog, Elsie.

E. N. Anderson is Professor of Anthropology, Emeritus, at the University of California, Riverside. He received his Ph.D. in Anthropology from the University of California, Berkeley, in 1967. He has done research on ethnobiology, cultural ecology, political ecology, and medical anthropology in several areas, especially Hong Kong, British Columbia, California, and the Yucatan Peninsula of Mexico. His books include *The Food of China* (1988), *Ecologies of the Heart* (1996), *The Pursuit of Ecotopia* (2010), *Caring for Place* (2014), *Everyone Eats* (2014), *Food and Environment in Early and Medieval China* (2014), and, with Barbara A. Anderson, *Warning Signs of Genocide* (2012). He has five children and five grandchildren. He lives in Riverside, California, with his wife Barbara Anderson and three dogs.

FIGURES

ACKNOWLEDGEMENTS

We appreciate the comments and suggestions of Barbara Anderson, Lynda Jean, William Jean, and the anonymous reviewers whose efforts improved the work. We also appreciate the work of Matthew Gibbons and Manas Roy at Routledge. Luke Wisner produced the maps.

PREFACE

The initial, tongue in cheek, title of this book was "We're Fu*ked," a name meant to convey the desperate situation we find ourselves in. But then we thought that this might be a bit pessimistic. We face a climate crisis but may still have some time to mitigate its impacts and to address all the other environmental problems we humans have created. But is there enough time? Is there time to do what is necessary to keep this traumatized ship afloat, or do we just get another bottle of champagne and listen to the band play "Nearer, My God, to Thee" as the ship goes under?

The situation is desperate. We can reduce carbon emissions, we can clean up the air, we can pick up the plastics, we can stabilize our farming system, we can deal with water shortages, we can do all this and more. That is, if we decide to. The will to act, not the ability to act, seems to be the roadblock. A few people are getting far too rich following the current path to want to change things. Politicians are making too much money in bribes and campaign contributions. They are all thinking about today and ignoring tomorrow. If that were not bad enough, we are on track to destroy thousands of small traditional societies and the knowledge they have that could help us.

Here we try to provide some basic information about the state of the environment and some of its major issues. We are not trying to present a comprehensive list of problems or solutions. Instead, our goal is to provide straightforward information on some of the major problems so that the nonscientists can understand the issues and what is at stake so that they can choose to make a difference. We take an empirical approach to describing these problems; that is, we present scientific evidence backed by hard data rather than hopeful denial, speculation, wishful thinking, a head in the sand, and downright lies. Real information is often hard to get to people who only listen to right-wing media and politicians and are told that science is bad, that truth is not real, and that climate change is a hoax.

We take an anthropological approach and present examples of past successes and failures. We also highlight examples of how traditional peoples, past and present, have dealt with these same issues. There are lessons to be learned from anthropology, but only if we listen to them.

It is necessary for the public to put people in power who will put the Earth first—people who have the will to maintain a livable planet. That is the real solution.

1

THE PLANET WE LIVE ON

*"**Earth First**: We'll trash the other planets later."* (author unknown to us)

Earth is one of eight planets (plus some moons) in our solar system; the only one so far known to sustain life. Thousands of other planets in other solar systems (called exoplanets) are now known, but none has yet been shown to host life, although there are promising candidates. Thus, so far as we know now, Earth is the only place we can live. Saltwater oceans and seas cover some 71 percent of the planet, and of the remaining 29 percent is land on seven continents (Figure 1.1), about half of which are deserts or mountains where few people live. That leaves about 14 percent of the surface of the Earth as "habitable" for the eight billion people that currently live here.

Given that the Earth is the only place we do live, or can live, it seems obvious that we need to take care of it. People talk about, or have heard about, the environment; this about the environment or that about the environment. Environmental issues, environmental problems, changing environment, global warming, climate change, and its ensuing crisis. We have all been exposed to these topics. But what does this mean? People wonder "How does this affect me"? or "Why should I be concerned"?

It is difficult to put this in perspective. Many environmental issues do not directly or immediately affect most people in the US. Thus, many of us do not really comprehend why they should be personally concerned. How does deforestation in Africa affect us here in the US? It is sad that sea level rise in Alaska has destroyed a Native American cemetery, but how does that affect someone's life here in Los Angeles? How do these issues affect their rent? We need to understand that what happens to anyone anywhere ultimately affects everyone everywhere.

DOI: 10.4324/9781003560326-1

FIGURE 1.1 World map showing continents, oceans, and major locations noted in the text: (A) Amazonia; (B) Aral Sea; (C) Aswan High Dam; (D) Bali; (E) Brazil; (F) Costa Rica; (G) The Grand Banks; (H) the Hopi area; (I) Indus River Valley; (J) Iran; (K) Lake Baikal; (L) Los Angeles River; (M) Maasai area; (N) Madagascar; (O) Maldives; (P) Maya area; (Q) Mexico City; (R) Mongolia; (S) Mosel Dam; (T) Nile Delta Dead Zone; (U) Palau; (V) Panama Canal; (W) Phoenix, Arizona; (X) Rapa Nui (aka Easter Island); (Y) Salton Sea; (Z) Singapore; (AA) Tasmania; (BB) Three Gorges Dam.

We are the proverbial frogs in the water with slowly rising temperature and do not (yet) realize we are about to boil.

Consider this. You live in an apartment building, and if a fire started in one of the other apartments, would you, sitting in the living room of your apartment watching TV, be concerned? Maybe you smell some smoke, but it is not very bothersome, and you do not see any flames. The belongings of whoever lives in that other apartment are being destroyed, but your stuff is OK. The structural integrity of the entire building is threatened, but your place seems fine, and the TV still works. The fire may reach you but not for a while, so you do nothing. Really? Any rational person would react at the first sign of fire and would immediately deal with the threat, knowing the likely outcome if they didn't. The Earth is on fire, so what are we going to do about it?

People in industrial states (e.g., the US, Europe, China, Japan, India, and others, grouped together as "Western") tend to hold the view that they are somehow separate from the environment and "nature"—meaning anything people did not make and are above it in some way. This view can be traced back to the Bible. In Genesis

1:28, the world and nature were created first, then "Man"—an entity separate from and superior to nature—was created and told to "subdue" the Earth and "have dominion" over it. Thus, Western philosophy includes the view that people should "conquer" nature. As a result, many people today believe that we must overcome nature and bend it to our will. However, in the completely different version of creation in Genesis 2, verse 15 says people were put into Eden, and thus the Earth, "to dress and keep it." This "stewardship" command has always provided Western societies with a choice: subdue or care. This remains a lively issue of discussion in religious communities.

Unfortunately, the "dominion over nature" view prevails in economic quarters. This view continues to permeate Western thought and action as we separate ourselves from our natural surroundings by artificially creating closed environments, including our homes, offices, and cars. Ironically, many corporate and government people have conveniently shifted their view on this matter, arguing now that human activity is part of nature and so changes in climate caused by humans (anthropogenic) are "natural" and thus not of concern. People are indeed natural organisms, but what we do is of great concern.

The problems of modern industrial states go back into deep time. The first states usually engaged in conquest and ruled on a large scale. Production was inefficient compared to today's standards. Much of it—especially the large-scale farms and estates of the feudal systems or slave plantations that supported the elites—soon fell to the lot of enslaved persons, war captives, and other unfortunates. Pressure for wise use and efficient, skilled production was minimal. Fortunately for human survival, small farms continued, and independent small farmers had to be skilled and efficient. They carefully calculate the costs and benefits, maximizing the survival of themselves and their families in the face of difficult environments.

Many non-Western societies, often with smaller populations, do not see people as being separate from nature, correctly seeing them as just a part of the system. They hold a more ecologically friendly philosophy of life than Westerners generally do. This attitude is based on solid, hardheaded common sense, often shored up by traditional religion and morality. Their view is that since we are part of the system we must manage it for survival.

These non-Western societies have less of an impact on their environment, partly because most have smaller populations and less access to complex and destructive technology such as bulldozers to clear forests. Given the right conditions and incentives, the argument goes, they would do as Westerners do. In support of this argument, one can point to the possible prehistoric destruction of the habitat on Rapa Nui (Easter Island), the deforestation of most of Europe between about 8,000 and 5,000 years ago, and the chronic problems of the environment in traditional China and India. In contrast to the argument, we have the great societies of Southeast Asia and pre-Columbian America, which supported millions of people for centuries with comparatively little damage to the environment.

We now recognize that humans and their societies are an integral part of the environment. Human activity affects the environment, which is then altered and, in turn, affects human activities. The shape and form of the environment is dependent on its history, a history that includes humans. Yet it is also important to realize that humans are not just another animal. Humans are self-aware, cooperative, technological, aggressive, and highly social, which makes their interactions with the environment more complex. We have a concept of *responsibility*, though we sometimes forget it.

Human activity has a wide range of impacts on the environment, from exceedingly minor to catastrophic. Today, human activities are having huge impacts on our very home, traumatizing the very planet on which we depend, threatening our lifestyles and ultimately even our existence.

While the climate crisis is the most urgent of our problems, we will also focus on other things, such as pollution, warfare, and shortages of fresh water, farmland, forests, and fish. Decline in availability of these, especially fresh water and farmland, will force enormous changes in the world economy in the very near future—within 50 years. Global climate change greatly increases the severity of the crunch. Understanding and dealing with these challenges is a daunting but essential task.

How Do We Know What We Know?

Knowledge is gained in many ways: indirectly through stories (oral narritives) or religion (faith), and directly through observation, measurement, and experimentation. This latter system is called empirical (measurable and repeatable) science. All societies have both systems of faith and empirical science, varying in importance depending on the society. Western societies also have both systems, but empirical science usually dominates. Western empirical science uses a particular method to learn things and is the basis for the complex technology and understanding of the environment upon which we depend.

In Western science, empirical data (evidence) are obtained. For example, the weight of an item can be measured in a system such as pounds or grams, and if it is reweighed, the result will be the same. Data are then compared to other data to form a hypothesis, a proposed relationship between the data. To be valid, a hypothesis must be testable. The scientist then conducts the test and determines if the test refutes or supports the hypothesis. If refuted, the hypothesis is rejected and new ones proposed. If supported, the hypothesis is accepted for the moment and subjected to further testing, and if still supported, the hypothesis becomes generally accepted (but always subject to more testing).

To deal with a complex problem, a series of generally accepted related hypotheses might be combined to form a model, a theoretical construct built to approximate reality. The overall model, along with its component hypotheses, can then be tested. If any of the hypotheses must be adjusted, the entire model will be accordingly adjusted. Every test and adjustment improves the model. An example of this dynamic can be seen in the continual updating of an approaching storm; the more recent the information on the storm, the better the predictions.

If models survive the requisite testing, they could be elevated into theories, explanations of complex phenomena that have been thoroughly tested and accepted by scientists. Ultimately, after even more testing and refinement, theories would be elevated into laws, such as the law of gravity. Even gravity is still tested, and it is known that the usual laws do not apply in all circumstances, such as at the atomic (quantum) level or around black holes in space. While there is still a great deal to learn about gravity, no one questions its existence or effect on their lives.

In essence, then, Western empirical science does not "prove" things (even criminal courts only require "beyond a reasonable doubt"); it only calculates probabilities based on the strength of the hypothesis, model, theory, or law. Science gives us the best method to examine a problem and to make informed decisions. The constant retesting of hypotheses and models and the rethinking of results is what makes science so powerful.

For example, until roughly 500 years ago, most people thought the Earth was flat, not an unreasonable hypothesis at the time. But evidence increasingly pointed to a new hypothesis: that the Earth was spherical (this idea is older than 500 years). This new hypothesis was confirmed early on by ships circumnavigating the Earth. Since then, massive and overwhelming evidence has confirmed the sphere theory. However, there are people who still refuse to believe it and continue to claim the Earth is flat. You would not want such a person on your jury.

Or think of it this way. When you see the weather news about an oncoming hurricane, they show you what are called "spaghetti tracks," a series of lines depicting the various predictions of the course of the storm (and looking a bit like spaghetti). Each of the tracks is based on slightly different assumptions, perhaps some differences in atmospheric measurements, and separate models of hurricane behavior. Obviously not all the tracks can be correct, but they are all pretty close to each other and headed in the same direction. As the storm gets closer, more precise data are obtained, the possible tracks are revaluated and adjusted, and storm warnings go out. In the end, the final path of the storm is close to the predictions. Once the storm has passed, the models are revised and improved for tracking the next storm. Science in action.

Science requires evidence and constantly adjusts as more evidence is gathered. The law of evolution is a case in point. The term evolution simply means change in a thing or system, and since all things change, all things evolve. Case closed. Change is in response to some pressure, such as disease, changes in food supplies, loss of habitat, and the like. Adapt or die (evolve or go extinct)! All systems are always under pressure of some sort and so are constantly adjusting to changes. As the author Louis L'Amour put it, "The Only Thing that Never Changes is the Fact that Everything Changes."

However, when most people think of evolution, they think of human biological evolution, often couched in the erroneous idea that humans evolved from apes.

This is not true; no scientist thinks this. Humans and apes do share a common ancestor deep in time, but they are not on the same evolutionary branch. Even the Pope understands this.

We have known about evolution for at least 12,000 years, ever since people began to domesticate plants and animals and so became farmers. These plants and animals were selectively bred to change their forms to emphasize traits people wanted, such as giving more milk or having bigger seeds. This was biological evolution directed by humans, and every farmer knew how this worked. What was not understood was how biological evolution worked in nature, and it was not until 1859 that Darwin explained this mechanism. There are still plenty of details we do not know about biological evolution, but we do know the basics, and it is all based on evidence.[1]

The anti-science people will say that since we do not have 100 percent proof of something, it must be false. This is, of course, a ridiculous conclusion. And these are the same people that rely on electrical power for their coffee, GPS to get to work, cell phones to communicate, TV to get their fake news, the doctor they see and the medicines they get, and all the other things developed by the very science they deny. You cannot use science when it is convenient, then deny that same science if it is inconvenient.

So, when do we decide to act on some threat? If there is a 1 percent chance something will kill us, should we do something? 10 percent? 20? With climate change, we are perhaps 95 percent certain, although specific details are less known. Is it finally time to act?

What is the Environment?

The Environment is in us, not outside of us. The trees are our lungs, the rivers our bloodstream ... and what you do to the environment, ultimately you do to yourself.

(Ian Sumerhalder)

The environment consists of the surroundings within which humans interact. This is a very broad definition and it can mean pretty much anything you want it to mean. An environment can be defined as a room, a pond, a valley, a continent, the planet, the solar system, or even the universe. However, the term is commonly used in reference to a specific issue, such as "this" is bad for the environment or "that" is bad for the environment, leaving people with no clear understanding of what the environment is and leading to confusion and misunderstanding.

Here, our focus is on the environment of the planet as a whole. The Earth's natural environment consists of both physical and biological parts. The physical environment includes landscapes, weather, climate, and solar radiation (ultimately the source of most energy), plus things like the air and water. It is common to define portions of the planetary environment by its physical attributes, such as the Arctic (cold), desert (hot), marine (ocean), or even urban.

The biological environment includes everything biological in origin: plants, animals, and microbes, both living and dead. It is common to define an environment based on similarities in its biology (called ecozones), such as a forest, grasslands, or meadow. It is also common to use combinations of physical and biological elements to define an environment, such as the Sahara Desert being defined by a combination of geography, geology, temperature, and the types of plants and animals living there. People (and all other organisms) interact with themselves, other organisms, and their physical environment in complex ways, in what is commonly called an ecosystem. The size and scale of ecosystems can vary depending on how and why they are defined. The largest ecosystem currently defined is called the biosphere and is the global environment and all of its smaller interacting ecosystems.

There is also a cultural environment, one created by human societies, that overlies the natural environment. This cultural environment includes politics, economics, racism, nationalism, attitudes, philosophy, and technology, among others. The cultural environment thus shapes how a society interacts with its natural environment, and the impact human societies have on the natural environment causes changes in the natural environment. Thus, all environments on Earth have been transformed by people, some more than others. Even so, a person cannot use the survival skills suitable for the Jasper National Park in Times Square, and vice versa. We need the relative wilderness of Jasper just as we need the cities, and they remain very different places, satisfying different needs.

The Earth's environment is also defined by broad time periods called "epochs." For example, the "ice ages" fall within the Pleistocene epoch. At the end of the last ice age, about 12,000 years ago, the environment became warmer and drier, and the Holocene epoch began. Some scientists believe that sometime about 1950, we entered a new epoch called the Anthropocene,[2] an era defined by the massive human changes to the environment.

Humanity has gone from consuming 70 percent of the world's annual resource production in 1960 to consuming 170 percent today—drawing down the resources laid down over time; humanity has also reduced total vegetation by 50 percent over the last 10,000 years, leaving a great deal of bare land.[3] Continuation of the current practices in industry and farming could well create so much climate change that all life on Earth would be menaced (this is why it is called a climate crisis). Fortunately, we know enough to combine current practices with traditional systems and develop a high-yield, sustainable future.[4] That is, IF we decide to.

Ultimately, all people, all societies, impact their environment at some level. The trauma inflicted is a matter of scale. Small populations with simple technologies do only minimal damage, while large populations with complex technologies can do horrific damage. For example, Neolithic farmers deforested much of Europe using stoneaxes, but it took thousands of years. Today, we can deforest vast tracts of land very quickly using D-8 bulldozers. Think of the Earth as the skin on your body. A small burn on your arm from a hot stove will heal quickly and will not impact your life. But if that burn was larger, say some 60 percent of your skin, your very life would be in danger.

Control

Everyone wants to be in control of our lives, our finances, our crops, and the like.[5] We make every effort to control everything we can. We dam rivers to control water. We use herbicides to control weeds. We build fences to control the movement of animals and people. We pray to deities to control (or at least try to influence) weather.

For the last 12,000 years or so, people have increasingly sought to control certain plants and animals. This process, called *domestication*, involves selective breeding to increase the size or the quantities and qualities of their seeds, roots, hair, or milk and to decrease their independence and aggressiveness.[6] With domestication came farming and herding, and we now control the very existence of many species that can no longer survive without us.

While domestication generally refers to control of plants and animals, a more expansive definition of domestication could mean control in a more general sense. All societies strive to exercise control over their environment and use a variety of methods to do so (whether these are effective is another matter). As a result, many people believe that they control their environment to some degree, and some think they have total control.

All societies will purposefully alter, manipulate, and change their environments to achieve a desired result. The scale of change depends on a variety of factors, including the reasons changes are made. Large-scale manipulation or alteration of landscapes, such as damming rivers and flooding valleys or clearing forests for cattle pasture, is a relatively recent phenomenon. Technology is a factor, as someone with a bulldozer can manipulate things faster and to a greater degree than someone with a stone axe.

Large-scale physical changes are generally planned and conducted for specific purposes. The results might be seen as advantageous, at least at the time, but may have long-term negative effects. For example, the clearing of the rainforest in the Amazon Basin (and other places) is profitable to the farmer or rancher for a few years but destroys the forest to the long-term detriment of the forest in general and ultimately to people as well.

Farming is another example. To plant a crop, land (often substantial acreage) must be cleared of its natural ecosystems so that crops, a substitute ecosystem, can be planted. This results in the replacement of many wild species with a few domesticated ones (a loss of biodiversity) and often considerable alteration of the ground surface through mechanisms such as leveling, plowing, or terracing.

Societies also strive to control their landscapes. Some modify landscapes in a relatively minor way, while others construct new landscapes to suit their needs. A classic example of this is contemporary cities, completely artificial urban landscapes and environments. The Chinese art of *feng shui* is another example.[7] In feng shui, things must be properly arranged, even in landscapes, to insure harmony. Still other people move materials great distances around the landscape for the construction of specific facilities. This began early: Neolithic people moved large stones hundreds of miles to construct Stonehenge in England.

Irrigation systems are common elements in constructed landscapes. Such systems can be very small to quite extensive, with many, even thousands, of miles of canals, ditches, and other facilities being built. Dams can flood large areas, creating lakes, swamps, and eliminating portions of rivers, streams, and valleys, such as Aswan in Egypt and Three Gorges in China. Resulting irrigation can transform landscapes from arid regions to farmland (but at a cost).

Other forms of landscape management are through the use of ritual and religion. Such activities include world renewal ceremonies, such as fertility rituals and even human sacrifice. The Mexica (the group that included the Aztec) conducted both of these practices to ensure that the Sun would continue to rise and the world would persist.

Another way to control and maintain the environment is through ritual stewardship. To the Indigenous Australians, the land and its resources were formed during the Dreamtime, the time before people.[8] Certain places are very special and contain power related to the Dreamtime. Tribal elders are responsible for the maintenance of those places, and failure to properly maintain them could result in catastrophe. As such, the landscapes were intensively managed through ritual. Similar responsibilities are set out in the Bible, where Adam was charged as the steward of the land (Genesis 2:15) and the consequences of bad land management were detailed (Isaiah 34:11).

Managing Resources

A "resource" is something that is used. Water, air, living space, and time are resources used by everyone. While food is also used by everyone, not everyone eats everything. For example, many societies, from Biblical Israel to modern China, consider many insects as food and a resource. But Westerners take the very different view that insects are not food but generally considered to be pests, animals to be killed. Every society has a unique outlook on the resources they use.

Everyone manages their resources in some manner, even if very poorly. We have already touched on the management of environments and landscapes, but all societies also manage individual resources. Some resources are passively managed, such as through ritual. Mostly, though, people actively manage resources in a variety of ways, such as farmers and ranchers managing their crops and animals, companies exercising economic control over commodities, or governments managing economies or militaries.

Sustainability

By "sustainable," we mean using no more than that a system can produce over a long time. Most systems are sustainable for a while, but for how long? You can pump water out of the ground until suddenly you can't. Contemporary Western industrialized farming systems are very productive but are not ultimately sustainable due to soil erosion, chemical pollution, and diminishing water supplies.

Population growth is a major issue in sustainability. The rate of growth has slowed impressively since the great population-growth scare of the 1960s,[9] but the world population is still growing. However, in some places, populations are in decline, with too many older people and too few younger ones, so the availability of labor could be a major issue in some economies. If this "demographic transition" to lower birth rates continues, the world population will peak around 11 billion people, give or take a billion, but that will still be a lot of people to feed. It will mean about three acres of land and only a third of an acre of cultivated land per person.

Western industrialized farming can accommodate more people for a while, but ultimately population growth will overwhelm it, especially since we are covering some of the best farmland with houses and people. Shanghai, Shenzhen, and other exploding Chinese cities now cover some of the richest and most productive farmland in the world that will be out of production for the indefinite future. The same is true of Los Angeles, Buenos Aires, and Mexico City, among others. Similarly, intensification of farming can only go so far before it leads to erosion, biodiversity loss, and eventual system collapse. The questions are how long it takes to do this and what adaptations can save the system. When the system collapses, what will replace it, and will that replacement be adequate for current needs?

Related to sustainability is resilience, the degree to which a system can cope with and recover from stresses. Think of a sports team. There are the starters and the reserves. If a starter is injured, a reserve must substitute in. If that reserve player is not very good, it means that the team is not very resilient and could lose the game. Once a system is stressed, it might muddle through to eventually return to its original state (e.g., the starter returning and the reserve player going back to the bench). Or the system might adopt transformative changes that transform the system into a new state (e.g., trading players to improve the team).

Holling and Gunderson classically described the resilience cycle: (1) an initial "alpha" reorganization when the system is recovering from a crash; (2) a rising r phase when conditions improve; (3) a K phase when the system is stable but stagnating and threatening decline; and (4) an "omega" phase when it collapses. Smaller cycles, such as the cycle of population of one species in one location, are nested in larger cycles, which are nested in still larger ones—a condition known as panarchy (think of Russian nesting dolls). Holling was originally writing about cycles—natural or forced—in fisheries and similarly fluctuating ecological systems, but the cycle also works for human societies.[10]

A 1972 MIT study[11] concluded that industrial society will collapse by 2040 based on "five major trends of global concern—accelerating industrialization, rapid population growth, widespread malnutrition, depletion of nonrenewable resources, and a deteriorating environment." A reanalysis of that 1972 model suggests we are "ahead of schedule" and well on our way to such a collapse and that it may happen before 2040 unless we change our trajectory of destruction.[12] Degroot and colleagues[13] counsel us not to scare people too much, to avoid the term "collapse," and to focus on how we can adapt. OK, but we should be scared. As Ozzie Osborne counseled us, we are "going off the rails on a crazy train."

Systems are constantly adapting to small changes, but what about a big change, like a sudden system collapse? When a system collapses, it enters a period of chaos and adjustment, eventually establishing a new system—a process in biology called punctuated equilibrium.[14] For example, if the Western industrialized farming system collapsed, food production would have to make major adjustments before it could become stable again. In the meantime, what sorts of disruption would we see (e.g., famine, refugees, warfare)? New and ancient farming methods may be used, new and ancient crops grown, people would likely have to eat much less meat and milk, and the like. Painful but doable.

One enormous problem for planners and economists is the fact that humans naturally see small immediate benefits as more important than huge future benefits—especially the far future.[15] This is technically known as a "steep discount rate." We discount the future more than might be rational. The problem is that, in the age-old saying, "eat, drink, and be merry, for tomorrow we may die." A hundred dollars now looks a lot more promising than a guarantee of a thousand dollars in two years, because one may be dead or rich, or the promised payment may fall through. In short, it makes sense for individuals to discount the future. Children are routinely tested for levels of self-control by seeing if they will take a marshmallow now instead of waiting for two, three, or four marshmallows in one or two hours, or a day. They learn to wait, but adults often unlearn that lesson. As politicians say, "there is nothing beyond the next election."

The case is totally different for nations and communities. They depend on continuity of the resource base—on some degree of sustainability. Turning the environment over to looters and exploiters is suicide in the long run. Trying to break people's short-term thinking and make them look to survival over decades is notoriously difficult.

Economic competition can do the same thing. Short-term, destructive use often brings quick profit. A forest that could provide millions of dollars in benefits over hundreds of years, but only a few hundred dollars per year, can be cut down and turned into pavement or worthless grassland for an immediate profit of a few thousand dollars. This seems like "economic growth" and "good return on investment" to the bankers, but to anyone taking a long view, it is obviously a catastrophe. The deforesters seem to be outcompeting the conservationists, and the net result is disastrous for all concerned.

Tradeoffs are often less simple. Much of the tropical forest of the world has been cut down in the last few decades to be replaced by oil palms. This preserves a forest of sorts and is highly profitable in the long term. On the other hand, it contributes to climate change: the original cut-down trees decay but the oil palms do not fully replace them. It also eliminates the benefits of the natural forest—biodiversity, timber, fruits and nuts, and other values. Large-scale and small-scale deforestation for food production goes on constantly and has for thousands of years. Judging when the tradeoffs are worthwhile and when the forests should be left standing is fiendishly difficult.[16]

This habit of thought has been catastrophic for resources that must be managed for the long term, like biodiversity, freshwater, and forests. Taking it all now and hoping the future will take care of itself is all too common. Thus, resources that could be managed for the long term are rapidly overexploited for short-term gain, leaving little left for the future. The problem is the future has now arrived, and we have almost used up everything.

Another problem built into the human condition is that once a problem is identified, people take some time and effort figuring out what to do about it. Meanwhile, the problem gets worse. Then the attempted resolution may no longer work or may need revision. The problem gets worse again. Moreover, a couple of new problems often show up while we are solving the latest one. Thus, problems are always running at least a bit ahead of solutions. Very few environmental problems are securely solved. Too often, the best we can hope for is slowing the rate of decline.

Another issue, especially serious in environmental issues, is the tendency of problems to start small and then slowly and insidiously grow bigger and bigger. Overhunting and overfishing routinely exhibit this scenario: a few settlers or locals take a few deer or fish, and nobody notices; as the population of settlers or locals slowly increases, the take gets bigger, but only gradually. No one realizes until too late how bad the situation is becoming, and when they do, there is resistance against stopping a "traditional" take. This also leads to what is known as "shifting baselines" or the "shifting baseline syndrome": people slowly and gradually become used to fewer and fewer and fewer deer, birds, and fish, and young people grow up without any idea of the riches that were lost.

Most dangerous of all, the problem was usually caused by people who were making money from causing it, and they strongly resist the loss of their livelihood. Overhunting, overfishing, overgrazing, deforestation, pollution, and environmental damage by mining all have had this history. Control of overhunting was eventually achieved in many nations through getting hunters to agree to limits in the face of losing all game animals otherwise. Overgrazing and deforestation have been harder to control, even with agreement by grazers and loggers that restrictions were necessary. Worst of all, now, is the global climate crisis because of the extreme power of the fossil fuel industry and its allies in construction and agribusiness.

Western Resource Management

Western management systems generally operate under different goals from Indigenous/traditional ones. Western management is more focused on immediate returns at the expense of long-term stability and is quite intensive, resulting in extensive alterations of landscapes. Traditional management systems are smaller in scale and involve less landscape alterations. Most are designed to produce a steady state, with reliable production of resources over time. For example, when European farmers colonized North America, they did not acknowledge the native management and claimed most of the landscapes were wild and untamed "pristine wilderness." In reality, however, the landscape was not wild at all but was a well-managed and highly productive environment. The Europeans interpreted the matrix surrounding the patches and corridors intensively used by the Indians as being unused (and so available for colonization) rather than as being used differently.

Western management, especially in the last 200 years, has been treating sustainable resources as wasting assets. We have cut down forests and paved over the ground they once covered. We have exterminated extremely valuable species, such as the Passenger Pigeon, a billion highly edible birds exterminated by overhunting. We have allowed formerly soil-covered lands to erode down to bare rock, where nothing will grow again for hundreds if not thousands of years. We have replaced diverse and productive ecosystems with much less diverse ones.

Also, the pressure has been to maximize throughput—consumption of natural resources and transforming them into goods to use, especially commodities for consumers to buy. Obviously, no system can keep increasing its resource consumption forever, and we are now at a limit.

Often, the change from sustainable use to overuse is slow and insidious, as with hunting and fishing. It always seems possible to put one more animal on the land without overgrazing. We have watched while California puts in housing tract after housing tract, lawn after lawn, until no conceivable water management system could supply enough water to keep the suburbs growing. It always seemed that the system could manage one more house.

This shows that there is a basic conflict between preserving natural resources and using them for human wants and needs. In some cases, there is simply no way to save a resource; any use destroys it. This is most clear in the case of fossil fuels; their only uses are to burn them for fuel or make them into plastics and chemicals. Some resources can be recycled, like metals, paper, and some plastics. Still others can be managed sustainably. Trees can be replanted, crops grown again, soil restored by organic farming and other techniques,[17] and animals saved and allowed to multiply. Traditional societies, interested in the steady state, thus focused on renewables. Modern industry focuses more on nonrenewables, especially fossil fuels.

One problem is that of the rich taking the benefits of the poor. This is as old as the Bible and the Epic of Gilgamesh, which speak of the problems of saving cedar forests, managing water, and keeping pastures from depletion. Later, Robin Hood became a folk hero by exercising what was once a person's right to shoot deer for survival and subsistence. During the later medieval period, the nobles increasingly took over the forests and prevented ordinary people from hunting. The nobles got luxury food; the peasants starved. The folklore of almost every long-standing empire and nation has equivalents to Robin Hood—heroes who "poached" wild resources to feed their communities.

Meanwhile, pollution of all kinds is steadily increasing.[18] The hallmark of industrial societies for the last 300 years has been measuring growth and good by the amount of material put through the system. Conversion of raw material to use-goods is widely considered progress, no matter how inefficient, wasteful, and destructive. The result is a vast amount of valuable and useful material thrown into the landfill, or, far worse, into the world's waters and soil.

A taste of the future is provided by the catastrophes of the late 19th and early 20th centuries, caused by climate crises that the world order of the time was incapable of handling. Colonialism, national rivalries, and the consequent bureaucratism and hostility paralyzed governments.[19]

In addition, urban living cuts people from nature. In the US, the *Los Angeles Times* informs us that "Many children now spend less than 30 minutes *per week* playing outdoors."[20] Declines in hunting and fishing continue. Fewer Americans visit national parks and forests. More and more Americans lurk in air conditioning and live either in apartments or in suburban houses ringed by unnatural lawns maintained by heavy pesticide use. Irrational and extreme fear of snakes (whose bite kills about three Americans a year—often members of "snake-handling" religious sects), spiders, coyotes, and other animals is hyped by the media, especially via horror films, so people avoid the outdoors and prevent children from playing there.[21]

Today, the people who suffer the worst are minority and Indigenous communities, who have little power and wealth and cannot often resist rich, powerful looters. The steady contraction of Native American reservations in the US still goes on. It was recently done on a far greater scale in Brazil under the populist Jair Bolsonaro, whose government not only opened tribal lands to non-Indians but turned a blind eye to large-scale looting and robbery of Indigenous resources. Similar doings are as old as history and no less common now than in most of the past. The high colonial period from 1500 to 1900 was a special case, though; whole continents were stolen outright. That period is mercifully gone, but local looting remains serious.

Much of this sort of damage involves more direct conflict: people battling other people (see Chapter 10). Wars, raids, expropriation, land theft, and other large-scale violent actions are devastating to ecosystems. They also force people to take the short-term view, sacrificing the future by overusing resources now. A devastating problem has been powerful but destructive and wasteful people taking resources from small, militarily frail groups that were managing their resources well. This is the economic story of most of colonialism, especially in the Western Hemisphere, and continues unabated today, with small Indigenous groups being looted, robbed, and even massacred. Recent examples have involved the Rohingya Muslims in Myanmar, the Native Americans of Brazil, and Canadian Indigenous people displaced for mining oil sands. At worst, resources are destroyed simply to crush the

Indigenous people who were using them; bison were virtually exterminated in the late 1800s by the US for this reason, simply to deny them to Native Americans.

The All Too Common "Tragedy of the Commons"

When resources are not owned or regulated and when anyone has access to them, they are called *common-pool resources*. People are always tempted in common-pool situations to go for short-term gain at the expense of long-term return. This results in the overexploitation or destruction of the resource, a situation called the "tragedy of the commons" (following Garrett Hardin's famous article of 1968[22]). If the resource is valuable, there is considerable pressure to exploit it as fast as possible to maximize short-term returns. If only one player took that approach, all the others would be forced to join in and get theirs as fast as they could before the resource was gone. Thus, a free-for-all develops.

Consider this simple example. At a child's birthday party, there is a piñata full of candy. All the children gather, and when the piñata is broken, all the candy falls to the ground. This abundant resource (the candy) can now be exploited, and the children scramble to get as much candy as they can as fast as they can before other children get it. The aggressive kids get a lot, others get less, and some get none (those are the children crying). As a result, the resource (the candy) is rapidly depleted, and nothing is left for tomorrow. The same thing happens with fish, forests, and any other valuable, unmanaged resource; it just takes a bit longer and has far more serious consequences.

Think of what happened in California after 1848. When gold was discovered, tens of thousands of fortune seekers descended on the soon to be state and literally tore it apart looking for gold. There was no regulation or law enforcement, just a free-for-all. The environmental damage was immense (and still visible), and the impact on California's Native American societies was cataclysmic.[23] This same basic process is currently ongoing in other areas, such as the Amazon.

Overexploitation of resources was all too common in the past as well. For example, while ancient people generally lacked the technology to overexploit deep water (pelagic) fish, they did, in some cases, overexploit resources closer to shore, such as shellfish. An example of this is overexploitation of oysters by past peoples along the coast of Florida.[24]

Those with weak political power—the poor (as in most traditional societies), the young, and above all the unborn future generations—are at an enormous disadvantage. So are downstream users—the upstream people get to pollute the river unless the downstream users can sue or show force. The powerful, the rich, and the upstream thus tend to win out, even if their use of the resource base is poor. Thus, no one can conserve the resource, no one has the authority to protect it, and each player must take as much as they can. This leads to destruction of the resource base because profits accrue to the most predatory users ("ya snooze, ya lose").

Where some try to preserve the resource, they get forced out of the competition because the users get immediate profits and so prosper, while the former forego immediate profits and so go out of business. If those (the rich, governments, and/or companies) with strictly short-term, cut-and-run interests have the power, they are apt to destroy the resource, and a much greater long-term payoff is sacrificed. Many tragedies of the commons can be cited, from the overlogging and overgrazing of public lands in the US to the destruction of the cod fisheries of the Grand Banks of New England to the devastation of the world's rainforests by mining, lumbering, ranching, and other activities. In Africa, outside interests take advantage of weak governments to conduct illegal and unregulated mining and woodcutting, often using child labor and leaving devastated landscapes in their wake. In some cases, these outside interests are Private Military Companies (PMCs) clandestinely working for foreign governments, such as the PMC Wagner mining gold and diamonds in Africa for the Russian government.

Fortunately, it is fairly easy to manage common-pool resources sustainably. It requires that some group be in charge, one that can limit the resource and the use of it. Elinor Ostrom won the Nobel Prize for Economics by countering Garrett Hardin's mournful observations with cases and a theory of successful commons management.[25] Hardin, a generous scholar and scientist, accepted this criticism early and wrote many follow-up articles describing such success.

Much of what looks like a competition between "greed" and "unselfishness" thus boils down to small but quick profits vs. larger but very slow return profits. Such conflicts can be resolved by negotiation and treaty-making.[26] Politics should be concerned with such matters. Unfortunately, political systems are too easily taken over by the rich, who then capture the resources for themselves without leaving much for others[27]—hence Robin Hood and his moral kin. Hence, also, increasing "deaths of despair," as noted by Anne Case and Angus Deaton.[28] These have become so common in resource-depleted areas of the world that even right-wingers have reformed on the basis of Case and Deaton's findings. In this and countless more subtle ways, inequality kills.[29]

Some conflicts, however, are more directly related to values. Wildlands are valued for their beauty, their recreational potential, and very often for their religious significance. Holy places in Jerusalem, Mecca, Istanbul, and elsewhere are not "natural," but are very much a part of the long-term environment. Sacred mountains, lakes, waterfalls, rivers, and other scenic features exist all over the world. The British Isles have a surprising number of "holy wells," associated with saints but probably dating to much earlier cults. Particular animals may be revered, as cows are in India, or reviled, as pigs are in Near Eastern religions.

Traditional societies commonly manage common-pool resources by making the resource a part of community morality and almost always a part of their religion. In the vast majority of successful management cases, the environment was sacred, animals and plants were spirit beings, or there were divine guards. Environmental management was a sacred duty, and "nature" was respected. Religions across the world have teachings on the need to conserve and manage resources.

The Image of Limited Good

George Foster described the concept of "limited good."[30] Many societies, especially traditional agrarian ones where opportunity is limited, develop the idea that resources are strictly limited and gaining something means taking it away from someone else. This is basically a matter of seeing the world as a zero-sum game. The limited-good worldview goes beyond that. Foster showed that the limited-good idea is often extended to love, care, and concern, though these are not necessarily limited.

Even in contemporary society, with its open opportunities, limited good thinking is rampant. Even conservatives realize it has gone too far. Thus, even the right-wing industrialist Charles Koch, not usually one to admit problems with competition, writes: "Research by Gallup and Todd Rose…has shown that about two-thirds of people think that most others see the world through a zero-sum lens. They also overwhelmingly assume that most people believe that one person's success means another person's loss. Yet when you ask Americans how *they* see success, 90 percent say they want personal fulfillment more than they want to be better than others, and that each person's success can benefit those around them."[31]

Most Americans realize at some level that the US is, or was until recently, more like a positive-sum game: people could work together to make everyone better off. For many, the world now seems to be a negative-sum game, in which almost everyone is getting worse off and can only survive by cutting others down so that they fall faster. This explains the brutal "populist" and fascist politics of many countries. Such politics consists of making others suffer in hopes that the dominant group can keep its power, declining slowly if at all.

Lies, Lies, and More Lies

> When the truth is found to be lies and all the joy within you dies.
>
> *(lyric from Somebody to Love by Jefferson Airplane)*

The above issues are bad enough, but they are, to some extent, built into the human condition. Many environmental crises today, however, are quite unnecessary. They are based on outright selfishness and lying by powerful interests. The competition gets really ugly when a highly profitable but thoroughly unsustainable extractive industry comes up against regulation. Then, anything is fair game, from murder (especially of Indigenous activists in Latin America) to massive lying. The "Big Lie" technique was perfected with Hitler and his minister of propaganda, Joseph Goebbels, a brilliant but insanely twisted and anti-Semitic man who refined the Big Lie method to a fine art. The idea is that if you tell a Big Lie loudly enough and often enough, people will believe it.

After the Nazis, among the next to pick on this technique was the tobacco industry, which claimed for decades that tobacco was perfectly safe. They eventually had to pay some compensation, but it was insignificant compared to the fact that by the late 20th century, as many as one-fourth of all human deaths worldwide were being ascribed to tobacco.

The environment got involved when fossil-fuel companies discovered the Big Lie technique. At first, they denied the harm of emissions and pollution, but then came the far more serious problem of general climate change. This was a truly existential threat to Big Oil, since dealing with it meant reducing and ultimately shutting down fossil fuel use. To combat this threat, Big Oil initiated a world-wide public relations campaign, often using the same public relations firms as the tobacco industry, to convince the public that climate change was either a hoax or not caused by carbon emissions. In support of this lie, they argued that since not every single scientist agreed, it must be wrong.

Big Lies often bring with them a different kind of lie, not directly related but highly useful: racism and religious bigotry. The governments or corporations in question often use the divide-and-rule strategy of setting blocs of citizens against each other, based on "race," nationality, language, religion, gender identity, or any-thing else they can exploit to divide the public and set groups against each other. The giant corporations will then piously identify themselves with the dominant majority—conservative Muslims in Iran and Saudi Arabia, rural ethnic Poles in Poland, "Whites" in the US, and so on.[32]

A look around the world shows that there is no inborn tendency for people to hate "the other" or be biased against any particular group. The problem is that any group that has been historically disliked, or even just made into a minority, can be and often is targeted by cynical politicians and resource extraction interests. They were demonized and identified with those who wanted to "block progress" by stopping destructive exploitation of natural resources. Again, we can trace this back to the oldest records of humanity. Groups in the way were demonized and suppressed in ancient Israel, Mesopotamia, China, and onward. The worst has been in the last 150 years, however, when governments have found it expedient to exter-minate whole ethnic groups—millions of people at a time—to consolidate power and wealth.[33]

The giant agribusiness interests also became involved in Big Lies. They use oil to fuel machinery and to produce agrochemicals from fertilizer to pesticides. They also massively deforest; deforestation for agriculture is the cause of 15–20 per-cent of climate change. Moreover, cattle produce vast quantities of methane, a major contributor to climate change. The agribusiness firms also wanted to deny the harms of pesticides, which cause frequent cancers and other pathologies in agricultural workers and often in consumers. Pesticides are now linked to reduced male fertility and birth rates.[34]

The lies turned really deadly when they merged into general attacks on science by lunatic fringe cranks: anti-vaxxers, anti-public-health activists, anti-birth-control crusaders, and propagandists for all manner of quack medicines, conspiracy theo-ries, and fringe products. A genuine holocaust was reached when the anti-vaxxers, herd immunity devotees, and anti-public-health libertarians combined during Donald Trump's presidency to shut down reasonable responses to the COVID-19 (COrona VIrus Disease 2019) pandemic. The story is told in some 600 pages by Jonathan Howard in his book *We Want Them Infected*.[35] At first, many doctors believed COVID was mild (except among the elderly). They thought it could be more or less left to itself, like the common cold. Avoiding extreme measures would have made sense had this been true. Alas, it was wrong. COVID killed millions worldwide and left long-term and often very serious damage in one-tenth of cases. The doctors opposing emergency response often refused to change their minds even as more information became available. The result was hundreds of thousands, possibly a million, unnecessary deaths. The US, with seven percent of the world's population, had one quarter of the COVID deaths, although some other countries were equally irresponsible. Moreover, further epidemics will find the US very poorly prepared to cope.

The result is that whole political parties, starting with the Republican Party in the US but extending to extremist left and right parties all over the world, are now ferociously anti-science. Governments acting not only to shut down science but to reverse science-based policies have taken firm control in several countries, such as Brazil under Bolsonaro (now replaced, fortunately) and Hungary under Viktor Orban.

This world of lies goes far beyond the inevitable conflicts between short-term and long-term, widespread but small benefits vs. very large but very narrowly con-centrated benefits, and the like. Lies are thoroughly preventable. They should be countered immediately at all levels. It should go without saying, but probably does not, that racism, gender bias, religious bigotry, and the like are whipped up by the very people who are doing the lying and the environmental and ecological damage.

More than one expert on genocide has compared the current situation with the Holocaust, not because the politics are similar (though they are in many countries) but because the pattern of indifference to death and the resulting millions of refu-gees are familiar. Mark Levene[36] sees the current century as one of "omnicide."

The problem is not "capitalism," or "socialism," or any ism. It is modern indus-trial society. Even "industry," per se, is not the problem. The problem, as pointed out by leading environmentalist Bill McKibben, is the idea that we can, and even should, continue to consume more and more material on a finite planet. Modern industrial society adds to this obviously impossible dream the idea that we need not even be efficient about it. "Economic growth," measured by how much income is received, does not take adequate account of the quantities of natural resources—of primary production—that are destroyed and wasted, ultimately impacting gov-ernmental systems and demography.[37] Ecosystem services are almost ignored in accounting, as pointed out especially by Gretchen Daily.[38] Traditional societies, whether hunter-gatherers or early farmers, had to save and economize. Modern industrial societies ignore the concept.

Importing and Exporting

Much of the damage takes place in countries dependent on exporting natural resources to richer countries. They pay the ecological price, get few or no benefits, and, perhaps worst of all, cannot fight the system—they have little political power on the worldwide stage and depend on richer countries buying their goods.[39] Brazil, for instance, is being ruined by exporting cattle and soybeans at the expense of its own forests and savannahs. In these countries, working to save the environment is deadly; environmental activists are murdered with the governments turning a blind eye. The world loses almost one activist a day.[40]

Inequality is the best predictor of agricultural expansion despite productivity-per-acre improvement.[41] Agriculture in these countries is increasingly dedicated to producing more luxuries for rich consumers or exports. Lucas Chancet nails it in a book title: *Unsustainable Inequalities*.[42] Increasing inequality may be the most deadly process ongoing.

Inefficient growth would be bad enough if resources were the only issue. Far worse is the cost in human lives. "Economic growth" is a religion that demands vast human sacrifices. Millions of people die every year from pollution, global climate change, starvation caused by loss of traditionally available resources, and other costs of doing business in a world where business is measured in "gross national product" rather than human benefits. Calls for more realistic accounting have been made, such as Joseph Stiglitz and colleagues in *Measuring What Counts*.[43] Surely, we should take planetary survival and human survival into account. But we don't.

Things have reached the point where the junk we have made now outweighs all living things put together. Emily Elhacham and colleagues calculated that there are only four gigatons of living animals, fully 900 of plants, and perhaps as many tons of microorganisms. Human buildings and infrastructure weigh 1,100. Plastics, alone, now weigh eight gigatons, more than all living things put together. Some 9.2 billion tons of plastic were produced between 1950 and 2017.[44] We have halved the worldwide tonnage of plants over the last 10,000 years. Recycling is done widely but is less than adequate and far from what it could be.[45]

"Greed" is clearly at work, but we may step back and ask: Greed for *what?* Do we really want the world's natural resources—and natural beauty—to be sacrificed for throw-away plastic and manicured lawns? Did anyone get to ask? What do people really want? Are they getting it? Or has "economic growth," defined as throughput of resources, taken on a monstrous life of its own?

Traditional Resource Management

Consider an alternative: managing resources as practiced in earlier times and still is practiced in areas that have escaped excessive industrial development. Most traditional societies have small populations and so do not require huge quantities of any particular resource. Thus, most traditional society's resource management is small-scale, for example, dealing with hundreds of acres of farmland rather than the millions of acres Western societies deal with.

More important to us now, traditional societies tend to use a broad array of resources, both domesticated and wild, many more than most Western societies, and so tend to have a large and diverse set of management practices tailored to each resource. Such an array of management practices can be very complex and require that people in those societies have extensive knowledge of the resource and their associated environments. Modern societies depend heavily on only ten crops and animals: wheat, corn (maize), rice, white and sweet potatoes, sugar, manioc, cattle, pigs, and chickens. The base gets narrower as minor crops fall away. Popularization of a few local crops, such as quinoa, has helped, but in general traditional agriculture uses a wider selection of species and varieties (there are thousands of varieties of potatoes, but we use just a few).

A good example of the management of farmland is that of the Hopi, a Native American society residing in northern Arizona. The Hopi rely on growing corn for the bulk of their food, but they also continue to use a variety of other resources, including gathering wild plants and hunting wild animals. The Hopi land is rather arid, generally not suitable for farming, at least not using Western methods, and drought is always a concern.[46]

To ensure a sufficient corn harvest every year, the Hopi use 24 varieties of corn and 23 varieties of beans, planted in a diversity of locations and different soil conditions. The planning for planting begins in late winter or early spring. Fields are prepared, fences are built to provide some shelter from the wind for the young plants, and dams and canals are built and repaired. The actual planting of crops takes place in May after the danger of frost has passed. Many field types are used, each planted with its own variety of corn. Permanent fields near reliable water sources (springs and seeps) are planted every year. Temporary fields are planted wherever rain had moistened the soil. If a stream overflow forms a patch of wet soil, crops will be planted. Sand dunes that had been rained on are planted with special types of long-rooted corn to take advantage of moisture trapped in the dune. Such single-use gardens are spread across the landscapes, and while the crops in them often fail, some do not. The investment in them is small, and their number and dispersion across the landscape serve to reduce general crop failure. This system has worked for the Hopi for more than 2,000 years!

Southeast Asia is a region that until the 20th century managed to build great societies with dense populations, sustainably using the environment without exterminating animal or plant species, and even enriching some ecological systems. Like most tropical regions, Southeast Asia had vast forests that were enriched in food trees, selective planting, sparing wild trees, or simply letting seeds fall and grow. By 1800, Southeast Asia supported millions of people without huge environmental damage. Health was a problem, but modern medicine has since solved that—unfortunately after much of the region had bought into the modern throughput economy. Some Indigenous American societies came close to Southeast Asia in supporting large populations without devastating the environment.[47] We need to learn the lessons here.

The long record of human society is very mixed. Modern industrial society is the worst in all history in terms of environmental damage but is also the best in terms of supporting people in safety and comfort (at least so far). It has even gone far into sustainability and success in some forms of environmental management: protecting reserves, managing some highly prized fisheries and forests, and developing sustainable energy production. Above all, people have learned how to deal with health issues from epidemics to sewage contamination of drinking water. Other societies have their own patterns of hits and misses. The Maya, otherwise extremely aware of environmental matters, were devastated by droughts (and probably warfare) sometime about 1,000 years ago. Imperial China, for example, conserved rice paddy land and some animal species, but did not deal well with ecological crises.[48] Even Southeast Asian societies did not handle health issues such as disease and sewage well. Thus, every past or traditional society, and modern societies too, can contribute to the future; none has the ultimate answer or the full story.[49]

Respect and Belief

The real heart of sustainable management in traditional societies, however, is not size or experience but a focus on respect and responsibility.[50] Individuals in those societies are trained to respect each other and the environment, including its specific components such as fish or trees. Children are taught from the very beginning to act responsibly toward others, both human and nonhuman. Usually, these concepts are part of the local religion and thus are held to be ordered by the gods or spirits from the day of creation onward. Such beliefs persuade individuals to act responsibly, even when not watched or monitored by other humans. Irresponsible use might anger the gods or spirits and dreadful consequences may result, such as a tree falling on anyone who kills a tree or famine following overhunting. But often simple respect and responsibility—learned very early—are enough.

These simple but compelling ideas are effective to an astonishing degree. In the Great Basin of the western US, Paiute and Shoshone hunters depended on bows made from straight-grained juniper. But juniper does not naturally grow with straight grain. Anthropologist Philip Wilke found that prehistoric hunters would find a young juniper tree and carefully prune and manage it to grow straight so that they could harvest a straight-grained bow stave from the tree. As junipers are slow growing, the process took decades just to get the first good bow stave. An additional bow stave could be harvested every decade or so. These "bow trees" were tightly managed for the long term and so remained productive for long periods, with people taking only as many staves as the tree could stand.[51] People had to be trustworthy, not taking too many staves over time.

In the Peruvian Andes at 14,000 feet, co-author Anderson (ENA) learned that the best (and sometimes the only good) firewood came from a bush called *llareta*

(*Azorella compacta*). It grows only half an inch or less a year. Yet it is cropped sustainably—people take only a year's growth—even though the plants occur in remote areas where no one watches.

These are extreme cases but can be matched widely in the world.[52] No one kills or cuts trees on the sacred mountains and groves of Tibet and neighboring realms.[53] Tiny creeks in western North America were full of fish in 1800, despite the ease with which all those fish could be caught, even with the simplest technologies. Even in the US, where respect for the environment is limited, littering is much reduced from the 1950s and 1960s. Laws have something to do with this, but most of the change is due to teaching respect for others and for the landscape.

In the 20th century, humanity tried rapid industrialization, socialism, strictly enforced laws, unenforced laws, religion, ideology, "modernization" ideology and practice, eliminating smallholders, agricultural reform to create more smallholders, moral reform, airlines, big dams, roads, democracy, autocracy, everything. Nothing worked except maximizing the number of ordinary people who profited from or wanted the desired change. Thus, reform-for-smallholders did work, as did fighting corruption. Concentration of power (and thus autocracy and autocratic socialism), and above all, bringing in giant firms from elsewhere, were extremely counterproductive.

In short, industrial society as we know it has reached a limit. There is enough soil, water, forest, grassland, and ocean left to give everyone a decent living, but not to maintain endless growth in resource consumption. There is also an imperative and existential need to deal with the climate crisis. This requires a new societal organization, one based on efficiency and wise use rather than endlessly increasing throughput. It must be based on respect for all rather than national and international discrimination and hierarchy.

An ultimate case of modern industrial destruction for profit is "planned obsolescence." Computers, cellphones, and appliances are made to wear out or become obsolete within a few years. Computers, especially, must be replaced with incredible frequency, even when they show no wear whatsoever, simply because they are out of date for the new software. The rapid shift in data storage methods has led, among other things, to incredible losses of data as old storage forms become obsolete and decay. Reel-to-reel tape, Hollerith cards, and CD disks have gone down in succession to decay or be lost, all too often without the data being transferred onto newer platforms.

A related annoyance is loss of ability to repair anything. Farmers have dealt with the extreme case of the John Deere company's making machines that cannot be serviced except by the company—which means ruin if a machine goes down at a critical time on a remote, isolated farm. Farmers protested and have won limited "rights of repair" from John Deere, but many other cases have not been resolved so well.

In all these cases, annoyance and even serious loss are bad enough, but the ultimate sufferer is the environment and the resource base. Countless tons of metal, plastic, and glass in the form of perfectly serviceable but obsolete appliances are now in the world's junkyards.

The design features of the best traditional systems could certainly be revived for the future and combined with successful features of modern industrial society. The first and most important idea widely (but, recall, never universally) shared in traditional societies was respect not only for other people (all of them) but also for the environment as a whole. This entails fairness: equality as a general rule, proportionality otherwise—each to get what they need. Children need less food but relatively more nutrients than older people, for example. Next came full cost-benefit analysis: the actual costs of production were taken into account rather than being swept under the rug and hidden as long as possible, as is common in modern societies and well-known even in many traditional ones. Next comes efficiency: don't waste anything if you can possibly find a use for it, and re-use everything you can. Also, use it all. Within living memory, every part of a pig was eaten, from nose to tail. Every herb and weed had some use, as is still true among groups like the Maya of southern Mexico. Don't overuse: there is no benefit in taking all a resource just to say you've taken it. Particularly hard for many moderns to understand is the idea of using most of what can be restored fastest: fast-growing trees and fast-growing fish (use sardines that live a year, not orange roughly that take over 80 years to reach fishable size). Within communities, the rule was always to share and to leave "enough and as good" for others (in the words of John Locke, 1975[1697]).[54]

China has adopted more environmentally friendly policies in the last ten years, though ecological ruin still goes on. Such policies have brought major improvements to the efficiency and sustainability of food production, carbon take up (mostly via reforestation in the north, south, and southeast), soil retention, water retention, biodiversity conservation, and other goods. The far north and south are doing best. The classic economic heartland in central-East China is doing worst because they already had spent huge sums of money on the high-industry, high-agrotech sectors.

Notes

1 Darwin, Charles. 1859. *On the Origin of Species by Means of Natural Selection, or the Preservation of Favoured Races in the Struggle for Life*.Oxford: H. Milford; Oxford University Press.

2 Wallenhorst, Nathanaël. 2023. *A Critical Theory for the Anthropocene*. Cham, Switzerland: SpringerNature.

3 Bradshaw, Corey, Paul R. Ehrlich, Andrew Beattie, Gerardo Ceballos, Eileen Crist, Joan Diamond, Rodolfo Dirzo, Anne Ehrlich, John Harte, Mary Ellen Harte, Graham Pyke, Peter Raven, William J. Ripple, Frédeérik Saltré, Christine Turnbull, Mathis Wackernagel, and Daniel T. Blumstein. 2021. "Underestimating the Chances of Avoiding a Ghastly Future." *Frontiers in Conservation Science* 13 Jan, doi.org/10.3389/fcosc.2020.615419. Also see Chew, Sing C. 2001. *World Ecological Degradation: Accumulation, Urbanization, and Deforestation, 3000 BC–AD 2000*. Walnut Creek, CA: AltaMira Press.

4 Turbet, Martin, et al. 2021. "Day-Night Cloud Asymmetry Prevents Oceans on Venus but Not on Earth." *Nature* 598:276–280; Hawken, Paul (ed.). 2017. *Drawdown: The Most Comprehensive Plan Ever Proposed to Reverse Global Warming*. New York: Penguin.

5 Langer, Ellen. 1983. *The Psychology of Control*. Beverly Hills, CA: Sage.

6 On domestication, see Clutton-Brock, Juliet. 2012. *Animals as Domesticates: A World View through History*. East Lansing: Michigan State University Press.

7 See Anderson, E. N. 1988. *The Food of China*. New Haven, CT: Yale University Press. Santos, Gonçalo. 2021. *Chinese Village Life Today: Building Families in an Age of Transition*. Seattle: University of Washington Press.

8 A particularly good introduction to Indigenous Australian thought is: Rose, Deborah. 2000. *Dingo Makes Us Human: Life and Land in an Australian Aboriginal Culture*. New York: Cambridge University Press; For an Indigenous view of landscape management, see Pascoe, Bruce. 2014. *Black Emu, Dark Seeds: Agriculture or Accident?* Broome, Washington: Magabala Books Aboriginal Corporation.

9 Ehrlich, Paul, and Anne Ehrlich. 1968. *The Population Bomb*. New York: Ballantine.

10 Holling, Crawford S., and Lance H. Gunderson. 2002 "Resilience and Adaptive Cycles." In *Panarchy: Understanding Transformations in Human and Natural Systems*, Crawford S. Holling, Lance H. Gunderson, and D. Ludwig (eds.), pp. 25–62. Washington, DC: Island Press. See also Ungar, Michael (ed.). 2021. *Multisystemic Resilience: Adaptation and Transformation in Contexts of Change*. New York: Oxford University Press.

11 Meadows, Donella H., Dennis L. Meadows, Jorgen Randers, and W. W. Behrens III. 1972. *The Limits to Growth: A Report for the Club of Rome's Project on the Predicament of Mankind*. New York: Universe Books.

12 Herrington, Gaya. 2022. *Five Insights for Avoiding Global Collapse: What a 50-Year-Old Model of the World Taught Me about a Way Forward for Us Today*. Basil, Switzerland, MDPI.

13 Degroot, Dagmar, et al. 2021. "Towards a Rigorous Understanding of Societal Responses to Climate Change." *Nature* 591(7851):539–550.

14 Eldredge, Niles, and Stephen J. Gould. 1972. "Punctuated Equilibria: An Alternative to Phyletic Gradualism." In *Models in Paleobiology*, T. J. M. Schopf (ed.), pp. 82–115. San Francisco, CA: Freeman Cooper.

15 Arrow, K., M. Cropper, C. Gollier, B. Groom, G. Heal, R. Newell, W. Nordhaus, R. Pindyck, W. Pizer, T. Portney, T. Sterner, R. S. J. Tol, and M. Weitzman. 2013. "Determining Benefits and Costs for Future Generations." *Science* 341:349–351; Kahneman, Daniel. 2011. *Thinking, Fast and Slow*. New York: Farrar, Straus and Giroux; Kahneman, Daniel, Paul Slovic, and Amos Tversky (eds.). 1982. *Judgment under Uncertainty: Heuristics and Biases*. Cambridge: Cambridge University Press.

16 The issues in the above paragraphs have been explored over the years. An older but still unbeatable classic is: Murphy, Earl. 1967. *Governing Nature*. Chicago: Quadrangle Books. For updating, see Braje, Todd J. 2024. *Understanding Imperiled Earth*. Washington, DC: Smithsonian Books; Berners-Lee, Mike 2021. *There Is No Planet B*. Cambridge: Cambridge University Press; Wellenberger, Peter S. 2020. *America's Environmental Crisis*. New York: Central Park South Publishing. Wallace-Wells, David. 2019. *The Uninhabitable Earth*. New York: Crown.

17 For example, see the Kiss the Ground organization.

18 Jarrige, François, and Thomas Le Roux. 2020. *The Contamination of the Earth: A History of Pollutions in the Industrial Age*. Tr. Janice Egan and Michael Egan. Cambridge: MIT Press.

19 Davis, Mike. 2002. *Late Victorian Holocausts: El Niño Famines and the Making of the Third World*. London: Verso.

20 Campbell, James. 2016. "Making Room for Nature." *Los Angeles Times*, July 29, A17.

21 Ferguson, Gary. 2014. "The Great Fear of the Great Outdoors." *Los Angeles Times*, Dec. 21, A27.

22 Hardin, Garrett. 1968. "The Tragedy of the Commons." *Science* 162:1243–1248. For Hardin's later correction to his overreach, after he learned how good commons management can be, see his chapter 1991. "The Tragedy of the *Unmanaged* Commons: Population and the Disguises of Providence." In *Commons without Tragedy*, Robert V. Andelson (ed.), pp. 162–185. Savage, MD: Barnes and Noble.

23 Castillo, Edward D. 1978. "The Impact of Euro-American Exploration and Settlement." In *Handbook of North American Indians, Vol. 8, California*, Robert F. Heizer (ed.), pp. 99–127. Washington, DC: Smithsonian Institution; Hurtado, Albert L. 1988. *Indian Survival on the California Frontier*. New Haven, CT: Yale University Press.

24 Savarese, Michael, Karen J. Walker, Shanna Stingu, William H. Marquardt, and Victor Thompson. 2016. "The Effects of Shellfish Harvesting by Aboriginal Inhabitants of Southwest Florida (USA) on Productivity of the Eastern Oyster: Implications for Estuarine Management and Restoration." *Anthropocene* 16:28–41.

25 Ostrom, Elinor. 1990. *Governing the Commons: The Evolution of Institutions for Collective Action*. New York: Cambridge University Press; Ostrom, Elinor. 2005. *Understanding Institutional Diversity*. Princeton, NJ: Princeton University Press; Ostrom, Elinor. 2009. "A General Framework for Analyzing Sustainability of Social-Ecological Systems." *Science* 325:419–422.

26 Barrett, Scott. 2003. *Environment and Statecraft: The Strategy of Environmental Treaty-making*. New York: Oxford University Press.

27 Piketty, Thomas. 2017. *Capital in the 21st Century*. Tr. Arthur Goldhammer. Cambridge, MA: Harvard University Press; Piketty, Thomas. 2020. *Capital and Ideology*. Cambridge, MA: Harvard University Press.

28 Case, Anne, and Angus Deaton. 2020. *Deaths of Despair and the Future of Capitalism*. Princeton, NJ: Princeton University Press. See also Angell, David. 2017. *The Death Gap: How Inequality Kills*. Chicago: University of Chicago Press.

29 Angell, David. 2017. *The Death Gap: How Inequality Kills*. Chicago: University of Chicago Press.

30 Foster, George. 1961. "Interpersonal Relations in Peasant Society." *Human Organization* 19:174–178; Foster, George. 1965. "Peasant Society and the Image of Limited Good." *American Anthropologist* 67:293–315.

31 Koch, Charles, with Brian Hooks. 2020. *Believe in People: Bottom-up Solutions for a Top-down World*. New York: St. Martin's Press.

32 For the whole unsavory link between subsidies, industry, and racism, see Anderson, E. N., and Barbara A. Anderson. 2022. *Sustaining Social Conflict*. Lanham, MD: Rowman & Littlefield; Bellant, Russ. 1990. *The Coors Connection*. Cambridge, MA: Political Research Associates; MacLean, Nancy. 2017. *Democracy in Chains: The Deep History of the Radical Right's Stealth Plan for America*. New York: Viking.

33 Again, this and what follows is documented in Anderson and Anderson, previous citation; and sources cited there.

34 Knapke, Eric T., Danielly de P. Magalhaes, Mohamed Aqiel Dalvie, Daniele Mandrioli, and Melissa J. Perry. 2022. "Environmental and Occupational Pesticide Exposure and Human Sperm Parameters: A Navigation Guide Review." *Toxicology* 465:153017.

35 Howard, Jonathan. 2023. *We Want Them Infected: How the Failed Quest for Herd Immunity Led Doctors to Embrace the Anti-Vaccine Movement and Blinded Americans to the Threat of COVID*. Hickory, NC: Redhawk Press, The Catawba Valley Community College Press.

36 Levene, Mark. 2022. "The Holocaust Paradigm as Paradoxical Imperative in the Century of Anthropogenic Omnicide." *Genocide Studies and Prevention* 16:76–100.

37 McKibben, Bill, and Kim Stanley Robinson. 2023. *Democracy in a Hotter Time: Climate Change and Democratic Transformation*. Boston: MIT Press.

38 Daily, Gretchen (ed.). 1997. *Nature's Services: Societal Dependence on Natural Ecosystems*. Washington, DC: Island; Daily, Gretchen C., and Mary Ruckelshaus. 2022. "25 Years of Valuing Ecosystems in Decisions." *Nature* 606:465–466.

39 Bunker, Stephen G., and Paul Ciccantell. 2005. *Globalization and the Race for Resources.* Baltimore, MD: Johns Hopkins University Press.

40 Butt, Nathalie, Frances Lambrick, Mary Menton, and Anna Renwick. 2019. "The Supply Chain of Violence." *Nature Sustainability* 2:742–747.

41 Ceddia, M. Graziano. 2019. "The Impact of Income, Land, and Wealth Inequality on Agricultural Expansion in Latin America." *Proceedings of the National Academy of Sciences* 116:2527–2532.

42 Chancet, Lucas. 2020. *Unsustainable Inequalities: Social Justice and the Environment.* Tr. Malcolm DeBevoise. Cambridge, MA: Belknap Press.

43 Stiglitz, Joseph E., Jean-Paul Fitoussi, and Martine Durand. 2019. *Measuring What Counts: The Global Movement for Well-being.* New York: New Press.

44 Elhacham, Emily, Liad Ben Uri, Jonathan Grozovski, Yinon M. Bar-On, and Ron Milo. 2020. "Global Human-Made Mass Exceeds All Living Biomass." *Nature* 588:442–444; Subramanian, Meera. 2022. "Can Nations Rein in Plastics Pollution?" *Nature* 611:650–653.

45 Geng, Yong, Joseph Sarkis, and Raimund Bleischwitz. 2019. "Globalize the Circular Economy." *Nature* 565:153–155.

46 Hack, John T. 1942. The Changing Physical Environment of the Hopi Indians of Arizona. *Papers of the Peabody Museum of American Archaeology and Ethnology* 35(1):1–85. On the Hopi, a very thought-provoking book is: Brandt, Richard B. 1954. *Hopi Ethics: A Theoretical Analysis.* Chicago: University of Chicago Press. Brandt, a philosopher rather than an anthropologist, was particularly alert to the land ethic. On the agriculture itself, a stunning synthesis is: Benson, Larry V. 2011. "Factors Controlling Pre-Columbian and Early Historic Maize Productivity in the American Southwest. Part 1: The Southern Colorado Plateau and Rio Grande Regions. Part 2: The Chaco Halo, Mesa Verde, Pajarito Plateau/Bandelier and Zuni Archaeological Regions." *Journal of Archaeological Method and Theory* 18:1–60, 61–109.

47 Flores, Bernardo M., and Carolina Levis. 2021. "Human-Food Feedback in Tropical Forests." *Science* 372:1146–1147.

48 Pei, Qing. 2021. *Climate Change Economics between Europe and China.* Cham, Switzerland: SpringerNature.

49 Diamond, Jared. 2012. *The World Until Yesterday: What Can We Learn from Traditional Societies?* New York: Viking Press.

50 Anderson, E. N., and Raymond Pierotti. 2022. *Respect and Responsibility among Pacific Coast Indigenous Nations: The World Raven Makes.* Cham, Switzerland: SpringerNature.

51 Wilke, Philip. 1988. "Bow Staves Harvested from Juniper Trees by Indians of Nevada." *Journal of California and Great Basin Anthropology* 10:3–31.

52 Sponsel, Leslie. 2012. *Spiritual Ecology: A Quiet Revolution.* Santa Barbara, CA: Praeger.

53 Sponsel, previous citation; Huber, Toni. 1999. *The Cult of Pure Crystal Mountain.* New York: Oxford University Press.

54 Locke, John. 1975 [1697]. *An Essay Concerning Human Understanding.* Ed/intro Peter H. Nidditch. Oxford: Oxford University Press; Locke, John. 1924 (1690). *Two Treatises on Government.* London: J. M. Dent. The line "enough and as good" is in the Second Treatise, on p. 133 of that edition.

2

THE LIFE AROUND US

Life is improbable, yet here it is, don't screw it up!

Life is incredible. We find life everywhere we look on this planet and suspect it exists on other planets and moons in our solar system. Life exists even in the most unsuspected places: in highly toxic mine tailings, in the deepest part of the ocean, around hydrothermal vents where the water temperature is above boiling, and within rocks many miles deep. Life takes many forms, the most recognizable being plants and animals, but also including fungi, bacteria, and perhaps viruses.

Plants are organisms that conduct photosynthesis, a process that takes water and nutrients and converts them into organic material using sunlight as an energy source. Most plants use chlorophyll and are green because of it. Plants are less diverse than animals but still total over 290,000 species of flowering plants[1] that produce seeds and fewer but still impressive numbers of conifers, mosses, and ferns. There are 73,000 species of trees, including perhaps 9,000 yet to be described.[2] There are over 25,000 members of the composite family (Asteraceae) and even more orchids.[3] Some 2,300 to 2,400 new plant species are discovered each year.

In addition to providing food for many animals, plants produce much of the oxygen in the air. Some plants are quite small, such as algae, while others are quite large, such as redwood trees. The largest-known single living organism on the planet is the Pando grove in Utah, a five-mile-long single multitrunked aspen tree with over 40,000 trunks but a single root system and having the appearance of a whole forest. A network of honey fungus (*Armillaria solidipes*) in eastern Oregon is also huge, covering some 2,400 acres.

Animals are organisms that eat some sort of organic material, breathe oxygen (in air or water), are mobile, and reproduce sexually (there are, of course, some exceptions

DOI: 10.4324/9781003560326-2

to these rules). Worldwide, there are about 1,500,000 known animal species, of which a bit more than half—751,000—are insects. Ants are numerous and diverse: there are about 20×10^{15} individuals, making up 12 megatons of dry carbon, which "exceeds the combined biomass of wild birds and mammals and is equivalent to ~20 percent of human biomass."[4] Animals are typically either invertebrates (e.g., insects and mollusks) or vertebrates, those with backbones and internal skeletons. Animals eat either plants (herbivores), other animals (carnivores), or both (omnivores).

Darimont and colleagues find that humans use over 15,000 of the world's 47,000 vertebrate species for food, medicine, pets, and other predatory purposes.[5] Taking wild animals as pets is ecologically destructive predation, not care. It is threatening even formerly common birds, like parrots and the white-rumped shama of Asia. It affects thousands of species, reducing their numbers by an average of 62 percent.[6]

About 40 percent of the 15,000 species used—12 percent of all vertebrates—are now at varying degrees of risk of extinction. Thirteen percent (four percent of all vertebrates) are endangered. Several species survive only in captivity and is the only way to preserve them. Of these, the European bison, California condor, and a few others have multiplied enough to be released back to the wild, while others have gone extinct waiting for a chance.[7]

Humans are omnivorous animals; we can eat anything organic except lignin and cellulose (grass and wood) or chitin (such as the shells of animals like crabs). There are many other social omnivores, but humans are unique. We are the only animals known to form complex plans, especially for distant futures. All life affects the environment in some manner, but people have the unique ability to transform or devastate their surroundings and even alter the climate of the planet. We are terraformers.

Humans are also predators, but we eat a much wider range of foods than other predators. In fact, humans compete substantially with most other predators and prey on virtually all the species regularly taken by them, from river fish to land rodents. Traditional small-scale societies have often lived for millennia without depleting their resources, only to see them wiped out by modern industrial society in just a few years. Pacific herring sustained heavy fishing by local people for at least 10,000 years, followed by more than 90 percent reduction in a few decades once white settlers discovered the resource.[8] The same was true for other resources, from salmon to cedar trees.

It is important to realize just how resilient life is. Humans must do incredibly destructive things over long-time frames to wreak so much damage. An impressive story of resilience comes from ENA's California files. In northern California there was an iron mine, the Richmond Mine at Iron Mountain. The ore was duly extracted beginning in the 1860s, and the mine was abandoned in 1963. Groundwater seeping into the mine leached so many minerals out of the rock that it became extremely acidic (pH 3.5). At first, the water from the mine was so deadly that it was absolutely sterile (and designated a polluted superfund site). Then, downstream from the mine, acid-tolerant bacteria colonized the stream. Slowly, slowly, they spread upstream by mutating and by hybridizing with fellow members of the genus *Leptospirillum*. While we who followed the story cheered more and more loudly, the bacteria grew on upstream and finally *colonized the mine*, covering the whole deadly work with bacterial films.[9] Life triumphant.

Diversity

Diversity is a measure of the differences in things. There are many millions of plant and animal species, each adapted to its own niche (in essence, which soil plants like or what animals eat) and habitat (where a species lives). Different regions have different types of plants and animals, from the cold Arctic to the warm rainforests; from shallow bay waters to the deep ocean.

One source of diversity is the evolution of resistance to being eaten.[10] Plants depend more on being poisonous than animals do because plants can't run away. Even animals can become poisonous to predators, as some insects are. But then the plant eaters evolve tolerance to the plant poisons and even concentrate plant-derived poisons in their own bodies to discourage predators. Monarch butterflies do this, acquiring toxins from milkweeds.

Many insects, and humans, have evolved tolerance to the glucosinolates in the mustard family; we love mustard, horseradish, rapini, arugula, and other plants made pungent by these chemicals, but most animals are poisoned by them. We can handle thyme, mint, and other herbs; the flavorings in them evolved to poison herbivores. The evolution of specific toxins and specific resistances is an important driver of biodiversity.[11]

Another source is soil biota, a major driver of diversity in forests. More generally, pests track organisms, especially long-lived ones like trees, and will focus on specific species, which will inhibit one species from forming dense stands, thus increasing diversity. Diverse forests support a greater diversity of animals. Diverse plants draw on different soil nutrients, support different insects, and create different niches, allowing much more packing of diversity into a given area.[12]

Humans are quite diverse as well, not so much biologically (we are all virtually identical genetically, if not in appearance), but in behavior. There are thousands of different societies in existence today and an unknown number in the past. Each of these societies has (had) its own unique set of behaviors. Within small traditional societies, people's behaviors are quite similar; all speak the same language, have the same religion, have the same basic marriage patterns, and the like. In larger societies, there is much more variation in behavior. There may be many languages and religions, differences in dress and foods, and different viewpoints on various issues. Western societies tend to be large and diverse, with immigrants forming a myriad of ethnic groups.

Immigration has long been an issue. People tend not to like intruders, and in the US, the anti-immigration movement has a long history, from being anti-Irish to anti-Italian, to anti-Chinese, anti-Eastern European, and now to anti-Hispanic and anti-Muslim. This is born from fear of losing "power" or of being replaced, or simply from outright racism. Historically, immigrant communities have eventually become integrated into an increasingly diverse American society. They often become anti-immigrant in their turn.

On the downside, Western society colonizes other societies, mostly for economic gain. While colonization has occurred in many places at many times, going back to ancient Assyria, Western society's version is particularly aggressive. This creates a series of ethical issues, such as the destruction of indigenous knowledge, including useful things like new medicines, fine art, morality, philosophy, and the like. Assuming those small societies survive, Western colonization forces them to be "Western" (see Chapter 4). Other colonization is similar. This costs too much human suffering and cultural loss to be continued.

Why Is Diversity a Good Thing?

People do not generally like change or new things. But think of this: what if we had to all eat exactly the same thing all the time, or if everyone had to be dressed the same, or have identical furniture in your house? Would that be good? No. We all appreciate the diversity in available foods and restaurants, in clothing options, in housing, in investments ("don't put all your eggs in one basket"), and in ideas to solve problems and develop new products. Diversity is strength and permits a wider range of adaptations. It also makes life more fun.

Look at diversity in a more serious sense. Americans derive about 30 percent of their total food supply from corn, mostly field corn (or dent corn) and some sweet corn. Field corn is used to feed animals that people eat plus to make a variety of other products, including corn oil, high-fructose corn syrup, and ethanol. Sweet corn is consumed by people directly. We grow vast quantities of corn, and much of the Midwestern US is one gigantic corn field. For the most part, we grow four types of corn and despite having been genetically modified (GMO) to resist pests and diseases, corn is not really very diverse genetically (recall that the Hopi have 24 varieties). This leaves US corn susceptible to a disease that could greatly impact production and put everyone at risk. Farmers and governments realize this, and there are now programs to genetically diversify crops to protect them (and us) from a possible devastating disease. In essence, we are trying to increase our options, like the Hopi!

If a corn disease causing havoc to humans seems a bit hyperbolic, recall the Irish potato famine between 1845 and 1852. The Irish and people across much of northern Europe were dependent on the potato crop, largely on only one variety of potato. When the potato blight destroyed the great majority of the crop, it caused a famine that killed about one million people in Ireland alone.[13] Another 2.5 million Irish immigrated to the US at about that time, mostly because of the famine. Abandoned farms from that period still dot the Irish countryside, and the population of Ireland is still below its 1845 levels. A tragic lesson on dependency and lack of diversity.

Biodiversity, and Losing It

Biodiversity is a measure of how many species inhabit a certain area. Biodiverse regions, such as Costa Rica (one of the most biodiverse places on the planet), foster the development of new organisms and the way living things adapt to change. Think of biodiversity as like a natural Research and Development (R&D) department, conducting inventive and innovative experiments so species can adapt to change or to develop new species. Environmental stress is a catalyst for new biology and the development of diversity, similar to warfare being a catalyst for the development of new weapons or ways to use them.

The incredible range of biodiversity in the world is rarely realized by nonbiologists. There are millions of species out there. Estimates of how many differ by orders of magnitude. We have barely begun to count the species of nematode worms, bacteria, and even insects. Whole new forms of life occasionally turn up. A tiny but utterly bizarre microbe that turned up in Nova Scotia a few years ago proved not even remotely close to any other known life-form (except for an obscure and misidentified earlier relative). It was named *Hemimastix kukwesjijk*, the latter word meaning "little hairy ogre" in the local Mikmaq language.[14]

This diversity goes far beyond anything explainable by ordinary single-gene mutation and natural selection. Much more dramatic changes occur from hybridization, doubling or multiplying of chromosomes (polyploidy), chromosome breakup and recombination, epigenetics, and the spooky barely known rearrangement and modeling of genes and chromosomes by RNA.[15] Ordinary bread wheat is a hybrid of three species of grass, uniting all their genomes in a huge hexaploid genetic system. Lager beer is the result of fusing a couple of yeast species into a hybrid.

Perhaps the most amazing change was the coming of mitochondria, originally the union of an alphaproteo bacterium with an archaea species from the Heimdallarchaeote group, possibly four billion years ago.[16] Mitochondria produce as much energy as an equally diminutive lightning bolt. It would be a very small lightning bolt, since mitochondria are only visible with high-power microscopes, but still that is an impressive amount of energy being produced and supplied by every cell in our bodies every second of the day. This allows complex organisms to arise. Without that energy, there would be no plants or animals, let alone energetic humans.

But microvariation is not unimportant either. Cells in the same Petri dish, treated the same way, still differentiate from each other over time.[17] Tiny differences in environment have epigenetic effects. Mutations occur. Over time, different cell lines arise and grow farther apart. "Identical" twins never are; they have the same genes, but small differences in womb environment and the like result in unique individuals. Small local cell differences show up as birthmarks that allow parents to tell twins apart. Then lifestyle and minor womb differences show up. ENA's identical twin nieces, now grown up, can no longer be confused, as they once were. They look and act quite differently.

Conversely, vitally important genes are protected against mutation by various genetic mechanisms, reducing biodiversity and conserving function.[18] This is one reason humans share many genes with algae and seaweeds—the genes code basic life processes. One of them codes cytochrome C, basic to life and identical in essentially every organism.

Local isolation produces diversity; species evolve on islands, in restricted environments, in cut-off areas separated by glaciers or huge rivers, and otherwise in isolation from competitors.[19] They may arise in the same broad region, but only rarely. Highly differentiated habitats and long-stable regions produce more species; the Cape region of South Africa has a wide range of topography in a unique, tiny area of Mediterranean-type climate that has been relatively stable for millions of years, hence an enormous species richness.[20]

Sociability evolves in complex ways. As biologist John Thompson puts it, "The web of life has not evolved toward more conflict or more cooperation; it has evolved toward ever more complex interplays between conflict and cooperation."[21] Life forms of all sorts evolve complex societies, where individuals routinely sacrifice themselves for their kin.

Both species and societies must either adapt (evolve) to changing conditions or go extinct. If the loss of Arctic ice means polar bears cannot hunt, they will go extinct. If a business was unable to adjust to a loss of in-person customers during COVID-19, it went extinct (out of business). A decrease or loss of biodiversity reduces the biological genetic pool and thus the overall ability to adapt. This can be seen in the destruction of rainforest from cutting trees for timber, mining, and pasture.

While there have been mass extinctions in the past due to natural changes in the environment and coupled with climate changes, even past societies had some share in wiping out large animals in North America (e.g., mammoths) and Australia. (Not all the lessons we learn from past societies are cheery ones; we learn from their mistakes too.) Today, humans are the cause of the ongoing mass extinction, with many species lost every day. This is especially true of rare island life forms.

More general and devastating is the combination of logging, agricultural expansion, urbanization, mining, and pollution that have spread like wildfire from industrial nations to the whole planet. Forest degradation, even without loss of total cover, leads to population crashes.[22]

Madagascar, the "eighth continent" of biodiversity with over 80 species of lemurs and countless other unique animals and plants, is rapidly losing its native flora and fauna as rampant and poorly controlled exploitation devastates it.[23] Valuable trees were cut and exported with no effective control. ENA observed this on a study tour with colleagues and students. These were uncontrolled burnings of forests for shifting agriculture, contrasting dramatically with the carefully controlled burns of Maya farming in Mexico. There was Wild-West-style sapphire mining with no controls or safety precautions; human moles tunneled into cutbanks, which sometimes collapsed, killing them. In Madagascar, humans are not much better protected than wildlife.

Consider birds. Worldwide, there are over 10,000 bird species, with more than 428 billion individuals, and some 9,700 of those species are in decline.[24] In eastern

North America, shorebirds have all but disappeared, with perhaps a 90 percent decline in population since 1800. Other species of birds have been much harder hit, with a 95 percent loss in some populations. Grassland birds and aerial foragers are hard hit. North America lost, by a very conservative count, three billion birds between 1970 and 2019.[25]

The reasons for the decline of bird populations are no secret. They illustrate what we said about problems in Chapter 1. Overhunting was the first crisis, but this was largely solved by the Migratory Bird Treaty of 1904 (ratified later through most of the hemisphere) and local conservation laws. But then the loss of wetlands and shores to drainage and pollution became a crisis. Loss of wetlands led to subsequent loss of food for the birds. Over 90 percent of America's wetlands are gone. The remainder are the great dumping grounds for sewage and toxic chemicals.[26]

Then even more problems accumulated. Cats kill at least 2.6 billion birds a year. Flying into buildings kills another 900 million.[27] These are examples of another common environmental problem: something that begins as a minor issue but expands as more and more cats, buildings, and so forth are added to the mix. "It doesn't hurt to take just one," but if the number of cats or people "taking just one" increases from one or two to a million to a hundred million, catastrophe is guaranteed—but it starts so slowly that nobody notices until too late.

The issue of "shifting baselines" is very clear here. In the 1940s, the Western Meadowlark occurred in tens of millions all over the western half of the United States and well into Canada and Mexico. It is now reduced by more than 90 percent, thanks to insecticides, heavy grazing (it nests on the ground in grasslands), and suburbanization. Yet almost no one seems to notice. Few remember the days when a meadowlark was seen on every 100 yards of fencerow or telephone line. Humans are adapted to deal with immediate, obvious problems, like a charging lion or a flood or fire; we are not well adapted to noticing and dealing with a very slowly building crisis. This is relevant to issues ranging from global warming to deforestation, as well as bird life.

The most serious issue impacting the loss of birds is the decline of their main food supply, insects. People competing with birds for food is yet another issue. We take so many food fish that seabirds are starving. When 2/3 of the food fish are taken, the birds fail to reproduce, and unfortunately, we have all too many documented cases of this.[28]

Further, diseases such as bird flu and West Nile virus have impacted already stressed bird populations. Hawaii's endemic birds, for instance, are rapidly losing out to avian malaria that came in with introduced bird species. Most native birds are now extinct, and the rest are largely confined to areas above 1,000 meters, where mosquitoes that carry malaria do not do well. The mosquitoes are quickly evolving tolerance for higher altitudes, which means certain doom to most native birds.

As if this were not enough, diclofenac and nimesulide, common medicines used for inflammation and many other conditions in humans and animals, proved deadly to Eurasian and African vultures who scavenged the carcasses of animals treated with those drugs (as vultures tend to do). Vultures declined 99 percent in India and are declining fast in Africa and Europe.[29]

The loss of birds takes us farther down the path, to the loss of insects. By far the greatest reason for the loss of insects is the saturation of land with deadly pesticides[30] in countless forms. Farmers' fields are sprayed from airplanes. Cities are sprayed on schedule. Perhaps the worst abuses are in family homes and gardens. When ENA moved into his present house, he found a shed in the back filled with an appalling rank of deadly poisons: rank after rank of cans, bottles, and boxes to kill mammals, snails, slugs, spiders, insects, weeds, molds, fungi, and bacteria; in short, almost every imaginable life form. These had wiped out the predators that normally control pests, so the pests were worse than ever. Deadly general toxins, called "rodenticides," though they kill not only rodents but everything else, saturate much of the environment; rats soon learn to avoid them, but they poison enough small animals to accumulate up the food chain and kill off predators, making the rat problem worse rather than better.

The best-studied insect (after the domestic honeybee) is probably the monarch butterfly. Iconic and easy to spot, it is carefully surveyed on its breeding, migration, and wintering grounds. Eastern North American populations "have declined by around 90 percent since the mid-1990s,"[31] when they were already in bad shape thanks to excessively intensive farming. There was a 22 percent decline in 2022 alone. The decline is due to "pesticides, climate change, loss of US grasslands, and illegal logging of their overwintering forests." Worst is that milkweed "has been devastated by increased herbicide spraying in conjunction with corn and soybean crops that have been genetically engineered to tolerate direct spraying. Overall, the western population is down more than 95 percent since the 1980s."[32] Neonicotinoid insecticides are particularly deadly to monarchs, as they are to bees, threatening the world with the loss of pollinators.

Still worse is the fate of amphibians. A fungus, *Batrachochytrium dendrobatidis*, has gone worldwide from its probable home in Australia, eliminating frogs as it goes; it has exterminated countless species, especially in the tropics of the Americas. A related fungus, *B. salamandrivorans,* is doing the same to salamanders. These fungi do best in cool, moist situations and thus are probably not related to climate change.[33] They spread with humans carrying infected materials. Many other fungi, from whitenose in bats to crayfish fungi in Europe, are rapidly wiping out species. Fungi and other organisms cause potato blight, wheat rust, and countless other diseases of food plants.

Added to this is the great shadow of the global climate crisis. The Earth is getting warmer and, in most areas, drier. Storms and other weather phenomena are getting worse. Deserts have become harsher, and many native desert species have become rare or have vanished, largely because it has become too hot and dry. Mountains lose their high-altitude cold habitats, eliminating species dependent on these. Animals can move north, but trees have more difficulty, though those with wind-distributed seeds may luck out. Life forms with considerable genetic diversity can manage so far—there is enough genetic variability to allow a few exceptionally heat-tolerant or fast-moving individuals to succeed, but rare species without wide ranges or great genetic variability are often doomed.

Another major factor in the loss of biodiversity is urbanization and human presence. Much of the best habitat in the US, and for that matter in the world, has turned to urban and suburban sprawl. People are about on the land, and native species have difficulty adapting to the loss of habitat and the omnipresence of people, dogs, cattle, and other animals. Even species that flourish with human contact cannot usually survive in inner cities or in heavily sprayed suburbs. This loss of natural habitat causes wild species into closer interaction with humans, creating conditions for the transmission of diseases from animals to people (zoonotic), such as was the case with COVID-19.

Multiply all this across the world. Westerners are fond of lamenting the decline of elephants in Africa, tigers in China, and whales in the Antarctic, but the appalling level of butchery of our own wildlife is equally damaging. We are losing not only beauty and diversity, but all the ecological services of a healthy community, from pest control to game and edible fish.

The result in terms of the sheer biomass of humans and wildlife is appalling. Greenspoon and colleagues recently calculated that wild land mammals are down to 20 million tons (3 kg per human), sea mammals (largely whales) to 40 million tons, as opposed to fully 390 million tons of people and 630 million tons of domestic animals. Much of the surviving wild mammal mass is deer and wild pigs. Few other mammals make a showing. So, humans and their tame animals weigh *seventeen times as much as all other mammals combined,* a shocking figure.[34]

A 2020 United Nations (UN) study reported that banks in 2019 loaned out $2.6 *trillion* worldwide for projects that destroyed wildlife and ecosystems and a similar amount into fossil fuel development.[35] Major logging and clearing of rainforest, big dam construction, and agribusiness projects are among the destroyers.

Shreya Dasgupta and colleagues assessed 105,732 plants and animals on the International Union for Conservation of Nature's Red List, the standard list of threatened and endangered species. They found literally no cases of improvement in category (as from "endangered" to "threatened"). They found that the status of many had dropped. As of 2019, when they wrote, over 28,000 are in the threatened or endangered categories. The situation is notably worse now, after only a few years.[36]

Hunting, often illegal (poaching), has long been a cause of decline in large animals. The world's rhinoceros species seem beyond hope; their horns are valuable enough to tempt poachers even in the face of death. Elephants are also losing ground rapidly. This trend is not new, as human hunters have been causing wildlife declines for many millennia.

For example, when people arrived in the Americas perhaps as early as 20,000 years ago, they encountered a suite of large animals (megafauna) that had never had to deal with human hunters and so probably lacked sufficient fear of people. At some point, people began to hunt these animals and to impact their populations. At about the same time, the climate warmed as the Ice Age waned, and these ecological changes stressed the animal populations toward extinction. The human hunters probably did the rest and the mammoths, mastodons, ground sloths, saber-toothed tigers, and other megafauna disappeared.[37] A similar story can be told about the large animals in Australia after humans arrived. Eventually people learned, and so did the animals; there were no major extinctions after the end of the last Ice Age about 10,000 years ago. Bison (aka buffalo) were driven over cliffs by Native hunters in large numbers, but only rarely, when the herds could support the toll, and the meat, hides, and bones were carefully saved and used.[38]

In addition to the illegal hunting of animals for trophies or contraband, wildlife is also illegally hunted for meat, generally in the forest or "bush," and this "bushmeat" is processed and sold in local markets. This hunting is generally done by local traditional people to supplement their food supply and income and is common in Africa, Asia, and Latin America. Virtually any animal seen, including primates such as monkeys, chimpanzees, and gorillas, will be killed and this has led to serious consequences to the populations of many species.[39]

In the African forests, the hunters are commonly farmers living outside the forest who then enter the region to find game. In other cases, the farmers have enlisted the traditional forest people to hunt and provide meat. This practice impacts the traditional forest groups since they must change their lifeways, diets, ethics (the bushmeat trade is illegal in many places), and traditions of game preservation and sustainable use. These changes ultimately endanger the survival of those forest societies.[40]

In addition, zoonotic diseases (e.g., HIV and Ebola) are more likely to be transmitted to the dense farming populations by infected forest animals. Such diseases have not only infected the local populations but have spread worldwide and killed tens of millions of people.[41]

A similar situation can be seen in the fur trade in northern North America between 1600 and 1850. The French and British developed a market for the furs of various animals and enlisted the Native Americans of the region to hunt/trap the fur-bearing animals for them. This economic activity had a number of tragic consequences. Many of the fur-bearing animals were hunted almost to extinction across the region (see the beaver example below), the populations of which are only now recovering and are today regulated by the government.

A second major consequence of the fur trade was the alteration of the Native societies. Prior to the fur trade, native people hunted for food but then changed to hunting for furs, meaning that Native groups became dependent on Europeans for food. Native people began to congregate around trading posts to obtain food (and, unfortunately, alcohol). The congregation of people made the transfer of diseases easier, and many people died as a result. Before the fur trade, Native people did not claim personal territory but, afterward, were forced to claim and defend their trap lines, resulting in an increase in interpersonal violence. Thus, men hunted fur animals while the women and children lived in towns around the trading posts, reliant on European foods and goods. Once the fur trade declined, the Native people were left with no furs and no food. A return to hunting for food was difficult due to the loss of skills over the previous generations, and there were few jobs to make money. The previously independent and prosperous Native groups were plunged into poverty, violence, and alcoholism and left dependent on government aid. These groups have still not recovered, although fur trapping remains important to Native peoples.[42]

Another species complex reduced or eliminated by humans is the beaver (*Castor fiber* in Eurasia, *C. canadensis* in North America). Its dams once stabilized the waters of millions of streams and rivers, creating local lakes, storing water, raising the water table, providing homes for wetland species, maintaining fish populations, and keeping the whole subarctic and temperate ecosystem alive. Beavers are masters of niche creation, a process long championed by Kenneth Laland and colleagues as a critical part of ecology.

Elimination of beavers because they interfered with drainage has been a problem, but the real massacre was for fur. In particular, a fad for beaver hats—formal hats made of felted beaver fur—led in the early 19th century to trapping out beaver over millions of square miles.[43] Companies even trapped all the beavers—leaving none to reproduce—simply to prevent other companies from moving into their territory (a tragedy of the commons). This happened, for instance, in the northwestern US.

The result has been failing water storage, loss of tens of millions of water birds, loss of fish, massive erosion, consequent loss of vegetation, and widespread drought. In the southwestern US, trapping followed by overgrazing and drought led to entrenchment of streams—they eroded canyons often 20 feet deep or more. The exposed local water tables drained away and led to desertification over vast areas.

Damage in Eurasia was much more gradual, taking place over centuries, but the same levels of deterioration must have occurred. Beavers have recently been reintroduced to England, where they had been exterminated, so places like Beverly—"Beaver Meadow"—will once again deserve their name. So much ruin just so a few men could wear funny hats.

Hunting is better controlled in the world now, but all countries have problems with poaching and market hunting. Chinese law enforcement has, in recent years, caught people who had illegally captured, and almost always killed, over three million land vertebrates, to say nothing of fish and insects. Birds were the hardest hit, but mammals, reptiles, and amphibians all took an awful beating. Some 25 percent of the species taken were of serious conservation concern.[44]

At least we know something about the extent of the problem, though a great deal of subsistence hunting and local shooting of pests goes below the radar. For most of the world's countries, we do not have such good records. Wild animals are routinely taken for food, legally or not, from the US to Indonesia.

Far more animals, however, are killed by habitat destruction. Wildfires, flooding, urbanization, draining wetlands, and other impacts have many times the impact of hunting. Even if they are not killed, animals are often mistreated. Congress passed the Preventing Animal Cruelty and Torture Act in 2019, making cruelty to animals a federal crime. Support was unanimous in both the House and Senate.

The Problems with Extinction, and How to Solve Them

Why is it important when species go extinct? Life is a complex web, and any change (e.g., a species going extinct) has a ripple effect on its surrounding species and ultimately to all life. While it is true that the extinction of a frog species in Costa Rica will not have any immediate effect on your cable TV bill, it ultimately diminishes all life.

Think of it this way (using a classic metaphor). If you are on an airliner flying across the Atlantic Ocean and you look out the window and notice a missing rivet on the wing, what would you do? There are thousands of rivets holding the plane together, and this is only one rivet—no big deal, right? Then you notice a second missing rivet, then a third, then a fourth. At what point do you panic in the belief that the plane is falling apart and you are about to crash? Species are metaphorical rivets holding the web of life together, and while it is natural for species to go extinct, the high rate of human-caused extinctions (loss of rivets) threatens us all.

One could also view the Earth as a single "living" entity (called the Gaia hypothesis[45]) with various life forms on it. This is comparable to the multitude of species (e.g., intestinal flora, skin mites, bacteria, viruses, even some fungi) that inhabit a healthy human body. People might be seen as parasites or disease organisms infecting the "body" of the planet and that Earth will respond in its own way to the trauma inflicted by people.

Longer-established populations learn to conserve. In an extremely important article, Stephen Beckerman and colleagues showed that South American tropical forest societies carefully conserve and manage game if they are stable, in control of their land, and fairly densely populated on the ground, while societies that have low population densities and high rates of mobility do not.[46] This common-sense finding is fairly obviously applicable in other areas. It certainly fits evidence from other small-scale societies worldwide.

Above all, people conserve what they want to conserve. Loving nature, caring for plants and animals, preserving what you need to have a good life, and taking charge of your well-being is as direct a way to live and flourish as anyone can devise. One problem with industrial society is that it is based on treating everything natural as "raw material" to be turned into commodities. Modern members of industrial communities conserve perfectly well when they care. An extremely instructive example comes from the world of sport hunters and fishers. Ducks Unlimited, Trout Unlimited, and their many imitators have not only preserved the items they want to hunt but gone out of their way to find the science behind habitat protection (see recent issues of *Ducks Unlimited* for the latest on wetlands conservation projects).

Another inevitable conflict concerns widespread but small benefits to everyone vs. very large benefits to a few. Consider the draining of small, temporary, or marginal wetlands. The benefits of maintaining lakes, rivers, and seas with minimal pollution and degradation are enormous and obvious, but even that does not prevent them from being damaged by developers. These were protected under the US Clean Water Act, but protection of them was stripped by the Supreme Court, basically because the interests of a few developers were very large and concentrated while the benefits to the general public tend to be long-term, diffuse, and hard to cost out. Ultimately, the vast mass of the public will lose, while the transient increase in developers' wealth will not last, but this does not make such conflicts any easier to resolve. Unfortunately, China has been harsher with wetlands, draining and cultivating or urbanizing most of its marshes, swamps, and lakes. Even today, despite better knowledge of ecology, a great deal of destruction of China's pathetic remnants of wetland still goes on.[47]

All these traumas are preventable. We can use integrated pest control to reduce pesticides to bearable levels. We can stop overhunting and poaching. We can establish reserves and enforce them. We can reforest. We can stop the spread of diseases. We can clean up pollution. The public is in favor of these actions—surveys consistently report this. We lack only political unity and direction.

We need to work with the local environment to preserve as much of it as possible and to create a system that maximizes benefits to humans with minimal damage to the rest of the biota. This cannot lead to supporting as many people at as high a level of material consumption as the industrial model but is obviously more sustainable. The extreme of this path is reached in tropical environments, where local societies have developed ways of using the tropical forest, cutting only a small amount at a time to produce staple food while depending heavily on fruit and nuts that use, augment, or replace the natural forest cover. In traditional Southeast Asia, rice took about 10 percent of the land; the rest was under forest, which ranged from wild rainforest in remote mountains to thoroughly managed forest of fruit, nut, and timber trees around settlements. The same pattern exists in traditional parts of tropical America today, where corn, root crops, and beans take up small fields, leaving much of the forest intact, though managed.[48] It was also rediscovered in many Polynesian islands as the way to rebuild after the initial ruin caused by rapid settlement, population buildup, and ecological overdraft. Like many other people in the world, the Polynesians learned from hard experience and rebuilt for sustainability.

Even less damage to the environment is done by most of the small-scale Indigenous societies of the world.[49] The cost is that they maintain small populations at low levels of material wealth. On the other hand, they manage resources well under often difficult conditions, and they do not destroy their environments. Often, in today's industrial world, they are the preservers of what local biota is left. This is true of pastoralists like the Maasai of Africa and the Mongols of Asia; of hunter-gatherers in the Kalahari; of small farmers in tropical America and Asia; and of local traditional farming communities in Europe.[50]

Nature as Therapy

Being outside in relatively natural or wild places or even in not-too-manicured gardens is famously therapeutic.[51] The healing power of these places has been known since ancient times. Medieval societies around the world used gardens to heal mental problems. This continues today. "Forest bathing" is now popular; this consists simply of being out in the forest, reveling in it. Nudity is unnecessary, though delightful when practical (but wear insect repellant!). Gardens and natural areas for schools are strongly recommended, but sadly not within the tiny budgets of most of the world's educational systems; training children in even the most necessary skills is not a priority, especially compared to war and to subsidizing giant firms. Often, school gardens depend on motivated donors and citizens. At least they are happening and known to be beneficial.

Notes

1 Ramírez-Barahona, Santiago, Hervé Sauquet, and Susana Magallón. 2020. "The Delayed and Geographically Heterogeneous Differentiation of Flowering Plant Families." *Nature Ecology & Evolution* 4:1232–1238.
2 Cazzolla Gatti, Roberto, et al. 2022. "The Number of Tree Species on Earth." *Proceedings of the National Academy of Sciences* 119:e2115329119.
3 Mabberley, David J. 2009. "Exploring Terra Incognita." *Science* 324:472; Mandel, Jennifer, et al. 2019. "A Fully Resolved Backbone Phylogeny Reveals Numerous Dispersals and Explosive Diversifications Throughout the History of Asteraceae." *Proceedings of the National Academy of Sciences* 116:14083–14088.
4 Schultheiss, Patrick, et al. 2022. "The Abundance, Distribution, and Biomass of Ants on Earth." *Proceedings of the National Academy of Sciences* 119:e2201550119.
5 Darimont, Chris T., Rob Cooke, Mathieu L. Bourbonnais, Heather M. Bryan, Stephanie M. Carlson, James A. Estes, Mauro Galetti, Taal Levi, Jessica L. MacLean, Iain McKechnie, Paul C. Paquet, and Boris Worm. 2023. "Human's Diverse Predatory Niche and Its Ecological Consequences." *Nature Communications* 6:article 609; Scheffers, Brett R., Brunno F. Oliveira, Ieuan Lamb, and David P. Edwards. 2019. "Global Wildlife Trade across the Tree of Life." *Science* 366:71–76.
6 Conniff, Richard. 2017. "Loved to Death." *Scientific American*, Oct.:40–45; Hughes, Liam J., et al. 2023. "Global Hotspots of Traded Phylogenetic and Functional Diversity." *Nature* 620:351–357.
7 Smith, Donal, et al. 2023. "Extinct in the Wild: The Precarious State of Earth's Most Threatened Group of Species." *Science* 379:794.
8 Thornton, Thomas, and Madonna L. Moss. 2021. *Herring and People of the North Pacific: Sustaining a Keystone Species.* Seattle: University of Washington Press.
9 Bao, Zhongwen, Carol J. Ptacek, and David W. Blowes. 2023. "Extracting Resources from Abandoned Mines." *Science* 381:731–732; Denef, Vincent J., and Jillian F. Banfield. 2012. "In Situ Evolutionary Rate Measurements Show Ecological Success of Recently Emerged Bacterial Hybrids." *Science* 336:462–466.
10 Walters, Dale. 2017. *Fortress Plant: How to Survive when Everything Wants to Eat You.* Oxford: Oxford University Press.

11 Kliebenstein, Daniel J. 2018. "Plant Nutrient Acquisition Entices Herbivore." *Science* 361:642–643.

12 Furey, George N., and David Tilman. 2021. "Plant Biodiversity and the Regeneration of Soil Fertility." *Proceedings of the National Academy of Sciences* 118:e2111321118.

13 Salaman, Redcliffe. 1949. *The History and Social Influence of the Potato*. Cambridge: Cambridge University Press.

14 Lax, Gordon, et al. 2018. "Hemimastigophora Is a Novel Supra-Kingdom-Level Lineage of Eukaryotes." *Nature* 564:410–414.

15 Shapiro, James A. 2023. "Evolution without Accidents." Why Did Darwin's 20th-Century Followers Get Evolution So Wrong? *Aeon Essays*, July 2023.

16 Lane, Nick, and William Martin. 2010. "The Energetics of Genome Complexity." *Nature* 467:929–934; Niedzwiedzka, Katarzyna Zaremba, et al. 2017. "Asgard Archaea Illuminate the Origin of Eukaryotic Cellular Complexity." *Nature* 541:353–358.

17 Pelkmans, Lucas. 2012. "Using Cell-to-Cell Variability—A New Era in Molecular Biology." *Science* 336:425–426.

18 Monroe, J. Grey, et al. 2022. "Mutation Bias Reflects Natural Selection in *Arabidopsis thaliana*." *Nature* 602:101–105.

19 Lamoreux, John F., John C. Morrison, Taylor H. Ricketts, David M. Olson, Eric Dinerstein, Meghan W. McKnight, and Herman H. Shugart. 2006. "Global Tests of Biodiversity Concordance and the Importance of Endemism." *Nature* 440:212–214.

20 Latimer, Andrew M., John A. Silander Jr., and Richard M. Cowling. 2005. "Neutral Ecological Theory Reveals Isolation and Rapid Speciation in a Biodiversity Hot Spot." *Science* 309:1722–1725.

21 Thompson, John N. 2013. *Relentless Evolution*. Chicago: University of Chicago Press.

22 Betts, Matthew G., et al. 2022. "Forest Degradation Drives Widespread Avian Habitat Degradation and Population Declines." *Nature Ecology and Evolution* 6:709–719.

23 Goodman, Steven M., and Jonathan P. Benstead (eds.). 2003. *The Natural History of Madagascar*. Chicago: University of Chicago Press.

24 Callaghan, Corey T., Shinichi Nakagawa, and William K. Cornwell. 2021. "Global Abundance for 9,700 Bird Species." *Proceedings of the National Academy of Sciences* 118:e2023170118.

25 Rosenberg, Kenneth V., Adriaan M. Dokter, Peter J. Blancher, John R. Sauer, Adam C. Smith, Paul A. Smith, Jessica C. Stanton, Arvind Punjabi, Laura Helft, Michael Parr, and Peter P. Marra. 2019. "Decline of the North American Avifauna." *Science* 366:120–124.

26 Smith, Paul A., Adam C. Smith, Brad Andres, Charles M. Francis, Brian Harrington, Christian Friis, R. L. Guy Morrison, Julie Paquet, Brad Winn, and Stephen Brown. 2023. "Accelerating Declines of North America's Shorebirds Signal the Needs for Urgent Conservation Action." *Ornithological Applications* 125:1–14.

27 Andrew-Gee, Eric. 2016. "Bird Populations in Steep Decline in North America, Study Shows." *Globe and Mail*, Sept. 14. https://www.theglobeandmail.com/technology/science/report-finds-north-american-skies-quieter-by-15-billion-fewer-birds/article31876053/?fbclid=IwAR1wK7tc0jMUFYW7EkOPZV8SIQ7f7vklVn1uIrMJRQlA4X4RcM_u1WpM56w.

28 Cury, Philippe, et al. 2011. "Global Seabird Response to Forage Fish Depletion-One-Third for the Birds." *Science* 334:1703–1706.

29 Stokstad, Eric. 2021. "Vultures Face New Toxic Threat." *Science* 373:1187.

30 Maggi, Federico, Fiona H. M. Tang, and Francesco N. Tublello. 2023. "Agricultural Pesticide Land Budget and River Discharge to Oceans." *Nature* 620:1013–1017. "About 3 Tg [3 million metric tons] of pesticides are used annually in agriculture to protect crops." Mostly herbicides. "Of the 0.94 Tg net annual pesticide input in 2015 used in this study, 82 percent is biologically degraded, 10 percent remains as residue in soil and 7.2 percent leaches below the root zone. Rivers receive 0.73 Gg of pesticides from their drainage at a rate of 10 to more than 100 kg yr^{-1} km^{-1}. By contrast to their fate in soil, only 1.1 percent of pesticides entering rivers are degraded along streams" so rivers get

very unsafe. Surprise, it's worst in the Amazon, La Plata, and rivers of China, India, and Southeast Asia. Glyphosate is the worst in soil/root zone and river discharge, but metam potassium (a soil disinfectant) is as bad in rivers.

31 Boyle, J. H., H. J. Dalgleish, and J. R. Pusey. 2019. "Monarch Butterfly and Milkweed Declines Substantially Predate the Use of GMO Crops." *Proceedings of the National Academy of Sciences* 116:3006–3011.

32 Endangered Earth. 2023. "Monarchs Remain in Trouble." *Endangered Earth* (Center for Biological Diversity magazine), Summer 2023:9.

33 Zipkin, Elise F., et al. 2020. "Tropical Snake Diversity Collapses after Widespread Amphibi an Loss." *Science* 367:814–816; Stegen, Gwij, et al. 2017. "Drivers of Salamander Extirpation Mediated by *Batrachochyrium salamandrivorans*." *Nature* 544:353–356; Chin, Gilbert, and Jake Yeston. 2008. "Frogs Leap to Extinction." *Science* 320:586.

34 Greenspoon, Lior, et al. 2023. "The Global Biomass of Wild Mammals." *Proceedings of the National Academy of Sciences* 120:e2204892120.

35 Greenfield, Patrick, and Phoebe Weston. 2020. "Banks Lent $2.6 tn Linked to Ecosystem and Wildlife Destruction in 2019—Report." *The Guardian*, Oct. 27, https://www.theguardian.com/environment/2020/oct/28/banks-lent-1-9tn-linked-to-ecosystem-and-wildlife-destruction-in-2019-report-aoe?CMP=twt_a-environment_b-gdneco&fbclid=IwAR2HMQXLft1CJQcIAXqOeu5jJy_Uc0cSbKZrATXauQa7wqwBpL7GRZReO14.

36 Dasgupta, Shreya. 2019. "From over 100,000 Species Assessments in IUCN Update, Zero Improvements." *Mongabay*, July 18, https://news.mongabay.com/2019/07/from-over-100000-species-assessments-in-iucn-update-zero-improvements/?fbclid=IwAR07fEw-XcdCWQk2K7a2cIqLGOBv4B-qrjvPjXmA_JwkeGbihh2ZjYA_ZUs.

37 Some researchers believe that human predation was the primary factor in the extinction of the megafauna (see Turvey, Samuel T., and Jennifer J. Crees. 2019. "Extinction in the Anthropocene." *Current Biology* 29(19):R982–R986) while others believe that climate change was the primary factor (see Mann, Daniel H., Pamela Groves, Benjamin V. Gaglioti, and Beth A. Shapiro. 2019. "Climate-Driven Ecological Stability as a Globally Shared Cause of Late Quaternary Megafaunal Extinctions: The Plaids and Stripes Hypothesis." *Biological Reviews* 94(1):328–352). For the Australian example, see Van Der Kaars, Sander, Gifford H. Miller, Chris S. M. Turney, Ellyn J. Cook, Dirk Nürnberg, Joachim Schönfeld, A. Peter Kershaw, and Scott J. Lehman. 2017. "Humans Rather than Climate the Primary Cause of Pleistocene Megafaunal Extinction in Australia." *Nature Communications* 8(1):1–7.

38 Anderson, E. N., and Raymond Pierotti. 2022. *Respect and Responsibility in Pacific Coast Indigenous Nations: The World Raven Makes*. Cham, Switzerland: Springer Nature.

39 Ripple, William J., Katharine Abernethy, Matthew G. Betts, Guillaume Chapron, Rodolfo Dirzo, Mauro Galetti, Taal Levi, Peter A. Lindsey, David W. Macdonald, Brian Machovina, Thomas M. Newsome, Carlos A. Peres, Arian D. Wallach, Christopher Wolf, and Hillary Young. 2016. "Bushmeat Hunting and Extinction Risk to the World's Mammals." *Royal Society Open Science* 3(10):160498.

40 Duda, Romain, Sandrine Gallois, and Victoria Reyes-García. 2018. "Ethnozoology of Bushmeat. Importance of Wildlife in Diet, Food Avoidances and Perception of Health among the Baka (Cameroon)." *Revue d'ethnoécologie* 14. doi: 10.1000/ethnoecologie.3976.

41 Wolfe, Nathan D., Walid Heneine, Jean K. Carr, Albert D. Garcia, Vedapuri Shanmugam, Ubald Tamoufe, Judith N. Torimiro, A. Tassy Prosser, Matthew LeBreton, Eitel Mpoudi-Ngole, Francine E. McCutchan, Deborah L. Birx, Thomas M. Folks, Donald S. Burke, and William M. Switzer. 2005. "Emergence of Unique Primate T-lymphotropic Viruses among Central African Bushmeat Hunters." *Proceedings of the National Academy of Sciences* 102(22):7994–7999.

42 Bone, Robert M. 2016. *The Canadian North: Issues and Challenges* (5th ed.). Oxford: Oxford University Press.

43 Goldfarb, Ben. 2018. *Eager: The Surprising, Secret Life of Beavers and Why They Matter*, White River Junction, VT: Chelsea Green Publishing; Morgan, Lewis H. 1868. *The American Beaver and His Works*. Philadelphia: J. B. Lippincott.

44 Liang, Dan, Xingli Giam, Sifan Hu, Liang Ma, and David S. Wilcove. 2023. "Assessing the Illegal Hunting of Native Wildlife in China." *Nature* 623:100–105.

45 Lovelock, J. E. 1972. "Gaia as Seen through the Atmosphere." *Atmospheric Environment* 6(8):579–580.

46 Beckerman, Stephen, Paul Valentine, and Elise Eller. 2002. "Conservation and Native Amazonians: Why Some Do and Some Don't." *Antropologica* 96:31–51.

47 On the history of forests in China, see: Elvin, Mark. 2004. *The Retreat of the Elephants: An Environmental History of China*. New Haven, CT: Yale University Press; Harrelll 2023, previous citation; Marks, Robert B. 2012. *China: Its Environment and History*. Lanham, MD: Rowman and Littlefield; Zhang, Meng. 2017. "Market-Oriented Reforestation: Secularization of Timberlands and Shareholding Practices in Southwest China, 1750–1900." *Late Imperial China* 38:109–152; 2021. *Timber and Forestry in Qing China*. Seattle: University of Washington Press.

48 Anderson, E. N. 2005. *Political Ecology of a Yucatec Maya Community*. Tucson: University of Arizona Press; Fedick, Scott (ed.). 1996. *The Managed Mosaic: Ancient Maya Agriculture and Resource Use*. Salt Lake City: University of Utah Press.

49 Anderson, E. N. 2014. *Caring for Place*. New York: Routledge.

50 See, for instance, this stunning account of traditional management in the Mediterranean Sea area: Grove, A. T., and Oliver Rackham. 2001. *The Nature of the Mediterranean World*. New Haven, CT: Yale University Press.

51 Marsh, Pauline, and Allison Williams (eds.). 2024. *Cultivated Therapeutic Landscapes: Gardening for Prevention, Restoration, and Equity*. London and New York: Routledge.

3
OUR CLIMATE CRISIS

"We're living in a powder keg and giving off sparks" (lyrics from Total Eclipse of the Heart by Bonnie Tyler)

Here we briefly discuss climate change and its resulting climate crisis, the most pressing issue of our time and one that is noted throughout this book. The importance and likely impact of climate change will affect all life on the planet. We are watching the beginning of a slow-motion cataclysmic train wreck of our own making.

This is not climate change's first rodeo but the first caused by humans. The Earth's climate has changed many times, often radically, over the last four and a half billion years. Almost a billion years ago, the whole Earth froze under an ice sheet. The asteroid that ended the Cretaceous period 65 million years ago caused sudden heating and fires and then cold from dust filling the air. This ended the Age of the Reptiles (dinosaurs) and hastened the rise of the mammals. About 55 million years ago, volcanoes spewed so much carbon dioxide (CO_2) into the air that the climate was warmer for 200,000 years. Over the last several million years, the Earth has gone through cycles of warm and cold periods, called the Ice Ages, the last one ending about 12,000 years ago. All of these were natural events.

Even in the Neolithic as early as 8,000 years ago, humans began to alter the levels of atmospheric carbon through the rise of agriculture, including rice paddies in eastern Asia, and the deforestation of Europe.[1] As the farmers cleared the land and the trees were cut down, the decay of the trees emitted carbon, and their absence prevented some carbon storage. In addition, the newly domesticated livestock began emitting methane. However, after about 1800, the burning of fossil fuels by humans began to dramatically increase the level of CO_2 in the atmosphere, the start of the process of a rapid human alteration of the climate. By the late 1800s, the

DOI: 10.4324/9781003560326-3

effects of this increase in CO_2 began to be noticed, and warnings on its potential impact were made.[2] By the early 1900s, it was clear that increasing atmospheric carbon would result in the warming of the planet and the alarm was raised. Nobody paid much attention.

More recently, this warming trend became known as global warming. While the warming result is global, its effects are not. Some places are becoming warmer, some colder, some wetter, and many drier. Storms are increasing in frequency, in intensity, in property damage, and in lost lives. Flooding is more frequent and massive. What were once "hundred-year" weather events now occur almost yearly. New records for heat are set annually (even daily!). People are dying from the hotter summers. Plants and animals are going extinct or moving their ranges. Rainfall is also affected. In general, wet areas are getting wetter, dry areas drier, but unpredictable and surprising shifts are occurring. The complex ocean circulation system that produces "El Niño: and "La Niña" years in the Pacific is being altered in ways hard to predict.[3]

Thus, what we are really dealing with is a variety of environmental shifts due to the changing climate, a phenomenon more accurately called climate change. Even a small rise in average global temperatures, say 2°C (5°F), could be catastrophic to human societies as they exist today. This is the climate crisis we now face.

Climate, essentially the long-term average of weather, is difficult to see daily. We know and expect that summers are warmer than winters; rain comes in certain seasons, and animals migrate at certain times. People are used to small variations in these patterns and so think little of them. But changes in these conditions are currently so rapid and obvious that almost everyone now notices them. The once warm summers are now hot and getting hotter each year. The winters have less snow. There is much less, or much more, rain. The western US was in a "thousand-year" drought for over a decade. People clearly noticed that!

Climate change will not "destroy" the planet; it will significantly alter it. There will be no "new normal" for people to adjust to since the climate will continue to react to a warming environment and thus continue to change for the foreseeable future. We will have to adapt to this continually changing climate and hope we can stabilize it before we are no longer able to adapt.

As the planet changes and population grows, resources will become less available, and competition for those resources will increase. At least a 10% reduction in natural biotic wealth will occur, even with planned (hoped) limits. The distribution of food and water will change, as farmland becomes scarcer due to overuse or drought. Such changes are already occurring, as can be seen by the influx of climate migrants into the US. Such migrations may well lead to an increase in violence and even outright warfare. In addition, people in the more climate-impacted areas of the US will move to other places in the US (this has already begun), stressing those destinations.[4] It would seem more efficient to deal with the causes of the migrations rather than the consequences of them. It is like continuing to clean up blood on the floor rather than stopping the bleeding in the first place.

As of now, the lack of urgent action on the climate crisis makes it appear that we industrialized folks are willing to sacrifice our coastal cities to sea level rise, endure (in lives and money) more powerful storms, accept the starvation of millions of people across the globe from drought, and radically change the biology of the planet as many species go extinct, all so we can continue to drive our cars, have our air conditioning, and eat hamburgers for a few more years. Sounds like a massive exaggeration? Nope. It is the current reality.

Planetary Flatulence

The primary cause of current climate change is not natural; it is due to human activity through the emission of an assortment of gases into the atmosphere. These gases trap heat in the atmosphere, much as a glass panel in a greenhouse traps heat and keeps the plants inside warm. Hence the terms "greenhouse gases" (GHGs) and the "greenhouse effect." The major GHG is carbon in the form of CO_2, accounting for about 74.4 (or more) percent of total GHG emissions, followed by methane at about 17.3 percent and nitrous oxide at about 6.2 percent, with other gases making up the rest. Interestingly, while not a GHG per se, water vapor also contributes to heat retention.

The physics and science of the greenhouse effect is complicated but well known. Still, refinements to this knowledge are constantly being made as new information is obtained. In essence, ultraviolet and other radiation from the Sun heat the surface of the Earth, and some of the radiation is reflected back into space. Much of the reflected heat takes the form of shorter infrared rays, and these are absorbed and re-radiated by the GHGs. The lower portion of the atmosphere, the troposphere, is dense and so contains most of the GHGs, and much of the heat is trapped in this layer. As GHG concentrations rise, the heat trapped in the troposphere increases. Conversely, the thinner air above the troposphere (the stratosphere) loses heat more rapidly (a process known as stratospheric cooling) and helps keep the heat near the surface.

The Culprit: Carbon

Some 99 percent of the GHGs emitted into the atmosphere are in the form of carbon-containing molecules. The main one, CO_2, makes up about 0.04 percent of the atmosphere and weighs some 220,000,000,000,000 tons. The quantities of GHGs emitted into the air are measured in tons, a measure compatible with the mass of the atmosphere. In 2021, humans collectively emitted around 50,000,000,000 tons of CO_2 into the air.

The concentration of GHGs in the atmosphere is measured in parts per million (ppm), and as of 2023, CO_2 concentration was about 419 ppm, as directly measured in the air. Using measurements from dated ice cores, we know that in 1800, CO_2 concentration was about 280 ppm, meaning we have increased the CO_2 in the atmosphere by some 66 percent in the last 232 years. This is only getting worse every day.

The Basic Carbon Cycle

Plants take in CO_2 from the air, run it through photosynthesis, then emit oxygen as a byproduct back into the atmosphere—oxygen that is critical to all life, including humans. Thus, the greater the number of plants there are, the greater the amount of CO_2 gets removed from the atmosphere and stored (sequestered) in plant tissues, and thus the more oxygen there is. Animals also take CO_2 into their tissues (recall that all life on Earth is carbon-based) and "store" it when they are alive. When the plants and animals die, they decompose, and their stored carbon is usually released back into the atmosphere.

However, in many cases, the dead plants and animals do not completely decay but "fossilize" to varying degrees. Some will completely fossilize to stone, but others will only partly fossilize, often becoming peat, coal, or oil. In these cases, the carbon in those dead plants and animals does not reenter the atmosphere but is stored in the partially fossilized remains. Through time, vast numbers of plants and animals have lived and died, with many of their remains forming coal and oil. The most famous of these times, from about 360 to 300 million years ago, is called the "Carboniferous Period" due to the huge quantities of coal and oil formed.

People have used peat and coal for fuel for many thousands of years. While the burning of peat and coal released the stored carbon back into the atmosphere, the quantities were initially very small and had negligible impact on the environment. But as human populations grew and the demand for "fossil fuels" increased, more and more coal was burned, and the effects of the release of that stored carbon back into the atmosphere became noticeable. When oil became an important fuel in the late 1800s, the use of fossil fuels accelerated, and the amount of previously stored carbon released back into the air dramatically increased.

About half of the emitted CO_2 is absorbed by plants and the oceans. Plant cover takes up 25 percent of the released CO_2 through photosynthesis. In addition to the well-known tropical forests, the dryland savannahs and forests take up a great deal, as do perennial grasslands, which store enormous amounts in their roots. Thus, replacing perennial grasslands by trees is a notably bad way to absorb GHGs since there is more stored carbon in the grass roots than can be taken up by most trees.[5] On the other hand, replacing shallow-rooted annual grasslands by brush or trees would improve our situation greatly. However, as forests and grasslands are destroyed, less carbon can be absorbed by plants. Further, the capacity of the ocean to absorb and store carbon is reaching its maximum. In addition, the high levels of carbon in the oceans have made the water more acidic, negatively impacting all manner of ocean life.

Archaeologists have a love-hate relationship with carbon. Carbon has three forms (isotopes). Carbon 12 is far and away the most common, but there is some carbon-13, and even a bit of carbon-14. Carbon-14 is radioactive, is included in the carbon taken up by living things, and decays at a known rate, which allows archaeologists to date archaeological materials. This is the love part of the relationship.

The hate part is that "old" (and so no longer radioactive) carbon-14 sometimes gets mixed into archaeological samples, making the dates older than they should be. There is also extra carbon-14 formed by the testing of hydrogen bombs in the atmosphere. This "bomb" carbon-14 makes the dates younger than they should be. Fortunately, archaeologists understand these issues and can deal with them.

Other GHGs

Methane is a naturally formed GHG, and even though emitted in much smaller quantities, its greenhouse effect is some 80 times greater than that of CO_2. There are considerable quantities of frozen methane under the seafloor and in permafrost. As the planet warms, these are melting, releasing massive amounts of methane.[6]

Methane is also emitted through the digestive processes in animals, with the 1.3 billion cows and eight billion people being major contributors. The real problem is the cattle—their fermentation process in digesting plant tissue generates enormous amounts of methane. Termites also produce a great deal of methane, a byproduct of cellulose digestion. Methane is also produced by various human activities. For example, methane is emitted from waterlogged rice fields by the burning of agricultural waste (and natural wildfires) and the decomposition of organic waste in landfills (some of which is captured to fuel power plants). In California, landfills are the worst source of methane, even more than cattle and fossil fuels.[7]

Methane is also a major byproduct of fossil fuel production, released during oil and gas extraction and called "fugitive emissions." These emissions could be greatly reduced just by better maintenance of the facilities. On the plus side, methane degrades fairly quickly in the air. This means that cutting methane emissions could have an immediate effect on the total impact of GHGs.

In addition to methane, oil production also produces natural gas. However, in many cases, this gas is just burned off (the fires one sees at the top of long pipes in oil fields). This verges on the insane—wasting our least polluting fossil fuel simply out of laziness—but it remains standard in some countries.

A result of our global economic system is massive international trade in commodities that produce GHGs. Brazil is the top exporter of cattle, soybeans, and other such commodities and is deforesting at an appalling rate to produce them. Indonesia comes second. The leading importer of such goods is China, but Europe and the US are also major importers,[8] and the issue of globalized GHG emissions complicates regulation. Not only does commodity production release GHGs; the huge cargo ships that transport the commodities around the world release as much pollution as millions of cars. Interestingly, the Maersk shipping company is planning to switch its ships to methanol (renewable wood alcohol) fuel, and there is one commercial cargo ship that has begun using sails to augment its diesel engines, going back to a clean and green technology used by ships for thousands of years. Both are steps in the right direction.

Militaries also lavishly emit GHGs into the air via planes, tanks, vehicles, bombing, and, of course, outright war.[9] The US has the largest military with some 750 bases in 80 countries and is the largest single consumer of oil in the world. It is ironic that the US military emits huge quantities of GHGs to protect the sources of fossil fuels that are creating the problem in the first place. On the plus side, the US Navy is experimenting with renewable fuels for its ships.

In addition, climate change has resulted in huge wildfires across the planet that release the carbon stored in plants back into the air. The world's expert on wildfire, Stephen Pyne, has chronicled the rise of burning and the increase in danger worldwide.[10] The increase in burned land is appalling, and the loss of life and property is tragic. As Pyne suggested, perhaps we are entering the Pyrocene, the Age of Fire.

Another huge source is the burning or simply oxidation and warming of peatlands and tundras. These release methane trapped over thousands of years of plant growth in situations where acidic (peat) or frozen (tundra) soils prevent the release of decay gases from plant materials. Today, with warming and with exploitation of peat, problems occur. Protecting peat is a high priority. A less pleasant and far less natural source is the worldwide garbage heap. Landfills are enormous sources of GHGs. Recycling and rapid burial of what cannot be recycled could alleviate this.[11]

The Emitters

By far the main culprit in GHG emissions is the burning of fossil fuels. Until recently, coal was the primary fuel for electrical generating plants and for many decades coal powered ships, trains, and homes. In the eastern US, heating oil is still widely used for home heating. The bulk of GHG emissions comes from the production of electricity and heating. The US is shifting electrical production fuel from coal to natural gas, but China and India are still building new coal plants. The next biggest emitter is the fossil fuels (gasoline and diesel fuel) used in transport (cars, trucks, trains, planes, and ships). This is followed by manufacturing and construction, buildings, industry, and smaller sources.[12] In 2023, humans collectively emitted around 38 billion tons of CO_2, a 40 percent increase from 1990. We now emit some 1,600 tons of GHGs into the air every second.

Other human activities also emit GHGs, such as the production of cement and plastics. The production of cement "releases as much carbon dioxide into the atmosphere each year as Europe's 300 million cars: 1.5 billion tons…"; the same source notes that "Plastics…production worldwide releases 400 million tons of greenhouse gases a year."[13] On the plus side, there are new processes to make cement from algae, cutting carbon emissions from regular concrete manufacturing by 90 percent. Recycling using electricity can also enormously reduce this problem.[14] There are also new processes to make biodegradable plastics.

The US oil and gas supply chain alone releases about a million tons of GHGs (CO_2 and methane) a year. Other oil producers release comparable amounts. In 2013, two-thirds of GHGs were produced by 90 corporations, essentially the giant fossil fuel corporations, including the national firms of Saudi Arabia, Venezuela, Norway, and China, plus a few giant manufacturing and agricultural conglomerates. This is likely still true. Finally, CO_2 is also emitted through natural means such as volcanoes. This is usually a trivial source, but a major eruption can inject an enormous quantity of CO_2. Some 55 million years ago, volcanoes injected enough CO_2 into the air to produce warming comparable to that today. Over a few hundred thousand years, plant growth and rock uptake of carbonates cleared this and cooled the planet again.

Your Carbon "Footprint"

We hear about the carbon footprint. What does that mean? Simply put, a carbon footprint is the amount of GHG emission an entity is responsible for. Every individual, company, industry, organization, city, and country has a carbon footprint. In traditional societies, the carbon footprint of the society or person is very small; they may exhale some CO_2 or cut down some vegetation that decays and emits some CO_2, but such quantities are almost unmeasurable.

The carbon footprint of people in Western societies is far larger. They use electricity usually generated by coal or natural gas fired power plants (nuclear, solar, and wind power has a much smaller footprint), they drive gas-powered cars to work (electric vehicles are for now still largely dependent on electricity from the coal and gas fired power plants), eat food usually produced by carbon-emitting farms, live in houses built from wood no longer taking in carbon, or concrete actively emitting CO_2. It is virtually impossible to live a Western lifestyle and be carbon neutral.

Your city or town also has a carbon footprint, using asphalt to cover roads, concrete to make sidewalks, and police driving around in cars. Add up the footprints of industries in the city, city activities, the city residents, and other activities and facilities, and you get a footprint for the city. Add together the cities in a state or province plus all the individuals, industries, and farms in rural settings, and you get that area's footprint. Add together all the states or provinces plus the other activities of the nation, such as cargo ships, air transportation, and military, and you get the nation's footprint. Add all that together, and you will begin to understand the massive carbon footprint of contemporary society.

Emissions can be mitigated by planting trees. They take in carbon and create shade to lower temperatures. It is also valuable to have efficient public transportation, so people will drive less. We also need to build green (renewable, e.g., nuclear, solar, and wind) power facilities and to increase efficiency in transportation (e.g., gas mileage in cars). We need to use less water, and the list goes on. Farming more efficiently with less decay of vegetation and less use of chemicals is a particularly strong need. Overuse of fossil fuels is the major problem, but politicians take money from the fossil fuel companies and so take no tangible action. Fossil fuel companies are the major contributors to conservative and right-wing politics in many countries.

We also need to recycle as it takes less energy to recycle materials than to make new materials (aluminum being an obvious example). Recycling is not new and has been used by humans for millions of years. In prehistory, stone from old tools would be reworked to make new and different tools as technology changed. For example, obsidian (a natural volcanic glass) was used to make large spear points but as the bow and arrow became dominant, less obsidian was needed to make the smaller arrow points and the spearpoints broken and discarded long ago were recycled to make the smaller arrow points, saving the costs of going to the quarry and transporting new stone back home.

Another example of prehistoric recycling comes from the American Southwest. Pueblo buildings were made from rock, mud, and pine timbers. If a building or room was abandoned, the timbers would be removed and reused in the construction of new buildings. This was far more efficient than travelling to the forest, felling a new tree, and transporting it back to the town. Ancient people were not stupid and understood cost-benefit analyses.

Truth and Consequences

So what? How does climate change affect me? The consequences of a changing climate are many, mainly stemming from increasing global temperatures. This heat not only makes the land and air hot (not good for people), but it also increases the available energy in the oceans and atmosphere for use by storms such as hurricanes and typhoons. This increasing energy is uneven, meaning that some places will receive extraordinary rains, such as those that battered the eastern Mediterranean in 2023, killing thousands in Libya. Bangladesh and Sudan were also battered by flooding, displacing many thousands of people from their destroyed farms and homes. Other places will receive less rain and enter drought.

Tornadoes, hurricanes, floods, droughts, and other violent weather events are also increasing rapidly. For example, in October 2023, Hurricane Otis hit the Pacific coast of Mexico and was the most powerful storm ever recorded in that area. Otis strengthened from a tropical storm to a Category 5 storm in one day; that's right, one day![15] People had little time to prepare, and many deaths and substantial property damage resulted.

Such events are causing rapid increases in economic damage. Many insurance companies now refuse to insure properties in the most climate-change-devastated parts of the US, leading to local anguish. Obviously, this is leading to unjust situations, in which the least fortunate are impacted but the rich can escape. Global climate justice is thus an increasing concern.[16] Speaking of economic damage, drought has lowered the level of Lake Gatun in Panama, significantly impacting the ability of the Panama Canal to operate its locks, making the transit times for ships much longer. This bottleneck is threatening global trade and prosperity.

Climate change also increases all manner of other risks. Mosquitoes, ticks, flies, and other vectors can range farther north, stay out later in fall, rise higher on mountains, and carry diseases as they go. Warm water breeds parasites and microbes.[17]

The climate crisis is already forcing many people to migrate. Island communities in the Mississippi River mouth are being displaced by rising sea levels and an eroding delta. Farmers in Bangladesh face saltwater invasion. Farmers in dry areas face drought, and in tropics face changes in rainfall regimes. Applied anthropologist Lawrence Palinkas provides a full account of the millions facing immediate need to migrate and tens of millions facing it in future.[18] Such a situation is playing out now with migrants from Central America seeking to enter the US.

The GHGs together could start a runaway feedback loop that would spiral out of control, leading to Earth's surface temperatures that would destroy most life.[19] This would occur if the seas heated to something close to boiling point. Such a result is not likely now, but a continued increase in GHG release could eventually cause it.

Coupled with carbon emissions is the pollution of air, land, and water. Everyone in the world has now been massively exposed to the usual roll of disease germs and dangerous chemicals and particulates. Most are aware of it and know that it is damaging to health and to the world ecosystem. The "forever chemicals" used in all manner of manufacturing are now present in most humans, with effects that are not at all understood.

The problem is that we are now dependent on climate stability for agriculture, health, and nations. The relatively slight warming of the planet from GHGs will cost trillions of dollars and threaten human survival.

Tell Me It Ain't So!

The Big Lie of climate change being a hoax is based on the fact that science works on probabilities of outcomes, meaning there is some uncertainty about the specific timing and consequences of climate change. This small uncertainty is grossly amplified by skeptics and cited as "proof" that climate change is false. Now, it appears that the flaw in climate science is one of timing; it is happening faster than most scientists thought it would.

The resulting worldwide campaign to deny climate change was orchestrated and funded by the giant oil and coal interests, most notably ExxonMobil and Koch Industries, and other "Carbon Barons" such as Shell, Chevron, and Peabody Coal. These corporations poured hundreds of millions of dollars, over 50 years, into denial, starting and funding whole organizations devoted to concealing truth and outright lying. They spent hundreds of millions on lobbying governments around the world, successfully fighting all meaningful attempts to deal with climate (and other) problems.[20] These companies are now on a PR campaign to convince us they are in the lead to develop green energy. Yeah, right. Only recently has any progress been made, but it is inadequate. In 2023, California sued the fossil fuel industry over their lies.

An extreme of dishonesty is reached in Texas, a state long dominated by the oil industry. The industry dictates to many school boards and to state education officials what will be taught, often writing the materials. Oilmen serve on many school boards.

Texas education officials convened teams of volunteers to rewrite the existing standards, and industry members volunteered for those writing teams and shaped the language around energy and climate. Industry members rallied to testify each time proposals to revise standards got a public hearing. When the board considered the rewritten standards for final approval, the industry appealed to members to advance their favored amendments.[21]

The fossil fuel industry worldwide tries and locally succeeds in operating similarly.

In the words of Jeffrey Sachs—an establishment economist not usually given to such strong language—

"Despite the urgent need for rapid decarbonization of energy, the process continues to be blocked by the corporate greed of companies such as ExxonMobil, Chevron, Gazprom, BHP, and the like. These companies will cause massive detriment to populations vulnerable to air pollution unless they are decisively brought under regulation, yet in the U.S., Australia, Canada, Russia, and other countries, the oil industry rather than the common good directs politics. We are in the political world of Machiavelli rather than Aristotle. The same tendency is clear in the food sector. Current global agricultural practices are wholly inadequate.... As with energy, these changes are...stalled or slowed by...food giants. These companies are massive lobbyists and major campaign contributors in the U.S."[22]

It should be noted that these corporations are very heavily subsidized by their governments and sometimes essentially *are* part of the government (as in Saudi Arabia and Brunei). They spend much of their huge subsidies on lobbying and on spinning disinformation. The appeals to "free market" by giant firms are completely disingenuous; the firms in question depend heavily on political policies of various kinds and on direct subsidy support.

Cutting these subsidies is the most obvious and necessary way to confront the climate crisis, but having carbon paid for by the firms that get the profits is important, as pointed out for 60 years now.[23] That would mean forcing the fossil fuel companies to clean up *all* the pollution they generate, cap old wells, repair neighborhoods to bring back their property values, pay the medical costs for the hundreds of millions of people whose health has been compromised by fossil fuel use, and, last but not the least, deal with climate change and pay the costs of it.

This would, of course, totally ruin the firms, showing that they have been uneconomic all along. The real price of fossil fuels has been paid by the rest of us—by all humanity and indeed by the whole Earth. It now comes to tens of trillions of dollars, and that is only the beginning.

Marco Grasso and Richard Heede[24] have recently calculated how much in reparations the companies should pay, year by year, from 2025 to 2050, to make up the minimal costs. Saudi Aramco, the Saudi Arabian government company, is the biggest producer, responsible for 53,714 million tons of CO_2 equivalent—4.78 percent of the total—since 1988. They should pay $42.7 billion a year. ExxonMobil comes in second, with 23,119 million tons (2.06 percent), and should pay $18.4 billion. Other companies owe progressively less.

This does not include paying for the enormous campaigns of disinformation managed by these corporations, which have devastated politics and political discourse, above all by normalizing massive lies and disinformation campaigns as standard politics.

A Lot of Hot Air

An obvious result of climate change is increasing temperature (although this is uneven), and as human-caused climate change drives temperatures higher, heatwaves are becoming more frequent and severe. The hottest month so far ever recorded was July 2024. Indeed, 2023 was the hottest year on record. Some 81 percent of humanity experienced at least one day of abnormal heat, and much of the world experienced days of record temperatures. High pressure over the western US and the Mediterranean sent temperatures well over 100°F for days on end.[25] The ocean water off Florida reached 101°F, an appalling temperature that guarantees the death of the coral reefs there. The future is already here.

Already, deaths of people from heat are rapidly becoming more common. They are more likely in inner cities, because of the heat island effect, and on farms, because of work in the blazing sun.[26] In 2022, some 63,000 Europeans may have died during that summer's heatwave,[27] and in 2023, heat waves in northern China and Texas killed hundreds more. In the furnace-like days of July 2024, people passed out from the heat and then suffered third-degree burns from falling onto the hot pavement.

Yet humans are one of the most heat-tolerant animals, protected by our sweating and other responses. The situation is far worse for small animals, and of course for "cold-blooded" animals that rely on the environment for heat and cannot control their temperatures well.[28] They are subject to rapid overheating and can be literally cooked alive. This is only going to get worse, but we have apparently decided to sacrifice tens (if not hundreds) of thousands of human and other lives and billions of taxpayer dollars to maintain corporate profits.

In Western industrialized societies, we are accustomed to dealing with the heat by using cooling systems, mostly refrigerated air conditioning (A/C). While these systems work well, they have some disadvantages. First, A/C systems use a great deal of electricity, mainly generated from power plants using the same fossil fuels that are a cause of the plant heating up in the first place. Second, A/C systems remove heat from the interior of a building and expel that heat into the air outside of the building, making the outside even hotter. Third, the refrigerant used is often itself a GHG. While A/C will cool your home, its use ultimately adds to the overall problem. Adjusting your thermostat up a few degrees would help.

The use of A/C is mostly dictated by the availability of electricity, a service most people in the US have. But the cost of electricity is increasing and is moving out of reach for poorer people even if available. Thus, it is the poor that will suffer the most from the heat. A cheaper alternative is evaporative cooling systems (e.g., swamp coolers), but they only work well in areas of low humidity.

When designing buildings, Western engineers generally automatically incorporate A/C to cool a building without thinking much about how to limit the heat in a building in the first place. Many homes in the US have black tar roofing materials or other materials that absorb heat. Elsewhere, we use black asphalt for parking lots and roads (some 70,000 square miles of which are in the US alone) that absorb heat. Not the best plan.

Some Progress?

In the past, people dealt with heat by building homes and other facilities with thick walls to provide insulation, by using small windows to limit interior sunlight, and by painting their buildings white to reflect the heat. These methods result in lowered indoor temperatures.

In addition, natural "air conditioning" has been used in the Middle East for millennia. These "wind-catcher" systems (bâdgir in Farsi) are based on using the wind to ventilate and cool a building. Rectangular chimney-like towers with openings on their sides (Figure 3.1) were built extending from the roof of a building to direct

FIGURE 3.1 A bâdgir (wind-catcher) in Yazd, Iran. Photo by Yahyadad (Wikimedia Commons, Public Domain).

the wind into the building for ventilation, sometimes using water or below-ground spaces to further cool the air. In some cases, ice could even be produced. Each building has a system designed specifically for it.

Western designers are beginning to consider such alternatives. Some buildings, new and old, are receiving reflective paint. False roofs are being built over some buildings so that the heat from the Sun does not directly hit the building. Where feasible, evaporative cooling is being used. Shades are being built over parking lots, often with solar panels on top, both to cool the asphalt and the parked cars and to generate electricity. Small bodies of water are being incorporated to cool the air and reflect heat. Trees and other vegetation are being added around buildings to cool the surrounding air through evaporation, to reflect heat, and to store carbon.

Cities are also beginning to change. Planting trees is an obvious and simple method to help cool a city. Reviving neglected rivers and riparian zones in cities is also useful. For example, the city of Los Angeles is reinventing the Los Angeles River, channelized and concreted over decades ago. The goal is to reestablish some natural water flow, create riparian habitats, reintroduce wildlife such as birds and fish, and build parks and wild zones. Considerable progress has been made.

Singapore has recently instituted a "Cooling Singapore" strategy aimed at lowering the temperature across the city. The emphasis is on vegetation: trees to create shade on rail routes, roads, pedestrian paths, and in parks. Other vegetation was placed on roofs and the sides of buildings to reflect sunlight and cool the buildings. New parks with trees and lakes were built to reduce air temperatures. Building materials that absorb less heat are being used and the shapes and placement of buildings are considered. Green (renewable) energy generation is also being integrated into the city. The program is being studied for possible use in other cities.

Notes

1 Kaplan, Jed O., Kristen M. Krumhardt, and Niklaus Zimmermann 2009. "The Prehistoric and Preindustrial Deforestation of Europe." *Quaternary Science Reviews* 28(27–28):3016–3034.
2 Abram, Nerilie J., Helen V. McGregor, Jessica E. Tierney, Mihael N. Evans, Nicholas P. McKay, and Darrell S. Kaufman. 2016. "Early Onset of Industrial-Era Warming across the Oceans and Continents." *Nature* 536:411–418. Also see Ruddiman, William F. 2003. "The Anthropogenic Greenhouse Era Began Thousands of Years Ago." *Climatic Change* 61(3):261–293.
3 Hwang, Yen-Ting, et al. 2024. "Contribution of Athropogenic Aerosols to Persistent La Niña-like Conditions in the Early 21st Century." *Proceedings of the National Academy of Sciences* 121:e2315124121.
4 Bastien-Olvera, B. A., et al. 2024. "Unequal Climate Impacts on Global Values of Natural Capital." *Nature* 625:722–727; Lustgarten, Abrahm. 2024. *On the Move: The Overheating Earth and the Uprooting of America*. New York: Farrar, Straus and Giroux.
5 Ahlström, Anders, et al. 2015. "The Dominant Role of Semi-Arid Ecosystems in the Trend and Variability of the Land CO_2 Sink." *Science* 348:895–899; Hanan, Niall P., and Anthony M. Swemmer. 2022. "Savannahs Store Carbon Despite Frequent Fires." *Nature* 603:395–396.

6 Crowther, T. W., et al. 2016. "Quantifying Global Soil Carbon Losses in Response to Warming." *Nature* 540:104–108.

7 Duran, Riley M., et al. 2019. "California's Methane Super-Emitters." *Nature* 575: 180–184.

8 Heng, Chaopeng, et al. 2022. "Land-Use Emissions Embodied in International Trade." *Science* 376:597–603.

9 Crawford, Meta C. 2022. *The Pentagon, Climate Change, and War*. Cambridge: MIT Press; Jorgensen, Andrew, et al. 2023. "Guns Vs. Climate: How Militarization Amplifies the Effect of Economic Growth on Carbon Emissions." *American Sociological Review* doi 10.1177/00031224231169790.

10 Pyne, Stephen J. 2021. *The Pyrocene: How We Created an Age of Fire, and What Happens Next*. Berkeley: University of California Press.

11 Girkin, Nicholas T., and Scott J. Davidson. 2024. "Protect Peatlands to Achieve Climate Goals." *Science* 383:490; Cosworth, Daniel H., et al. 2024. "Quantifying Methane Emissions from United States Landfills." *Science* 383:1499–1504.

12 Ritchie, Hannah, Pablo Rosado, and Max Roser. 2020. "Emissions by Sector." Published online at OurWorldInData.org. Retrieved from: https://ourworldindata.org/emissions-by-sector [Online Resource].

13 Geng, Yong, Joseph Sarkis, and Raimund Bleischwitz. 2019. "Globalize the Circular Economy." *Nature* 565:153–155.

14 Dunant, Cyrille F., Shiju Joseph, Rohit Prajapati, and Julian M. Allwood. 2024. "Electric Recycling of Portland Cement at Scale." *Nature* 629:1055–1061.

15 Garner, Andra. 2023. "Hurricane Otis Is a Deadly warning about Our Heating Oceans." *Los Angeles Times*, Nov. 10, A11.

16 Coronese, Matteo, et al. 2019. "Evidence for Sharp Increase in the Economic Damages of Extreme Natural Disasters." *Proceedings of the National Academy of Sciences* 116:21450–21455; Cripps, Elizabeth. 2022. *What Climate Justice Means and Why We Should Care*. New York: Bloomsbury Continuum.

17 Lemery, Jay, Kim Knowlton, and Cecilia Sorensen (eds.). 2021. *Global Climate Change and Human Health: From Science to Practice*. (2nd ed.). Hoboken, NJ: Jossey-Bass.

18 Palinkas, Lawrence. 2020. *Global Climate Change, Population Displacement, and Public Health: The Next Wave of Migration*. Cham, Switzerland: SpringerNature.

19 Turbet, Martin, et al. 2021. "Day-Night Cloud Asymmetry Prevent Oceans on Venus but Not on Earth." *Nature* 598:276–280.

20 These are serious charges, but they are very well documented. See Anderson, E. N., and Barbara A. Anderson, *Sustaining Social Conflict*, Lanham, MD: Rowman & Littlefield. Michaels, David. 2008. *Doubt Is Their Product: How Industry's Assault on Science Threatens Your Health*. New York: Oxford University Press, and Michaels, David. 2020. *The Triumph of Doubt: Dark Money and the Science of Deception*. New York: Oxford University Press; Oreskes, Naomi, and Erik M. Conway. 2010. *Merchants of Doubt: How a Handful of Scientists Obscured the Truth on Issues from Tobacco Smoke to Global Warming*. New York: Bloomsbury Press; Speth, James Gustave. 2021. *They Knew: The US Federal Government's Fifty-Year Role in Causing the Climate Crisis*. Cambridge: MIT Press.

21 Worth, Katie. 2022. "Climate Miseducation." *Scientific American*, July, 42–49.

22 Sachs, Jeffrey. 2020. "Sustainable Development Goals and Health: Toward a Revolution in Values." In *Health of People, Health of Planet and Our Responsibility: Climate Change, Air Pollution and Health*, Wael Al-Delaimy, V. Ramathathan, and Marcelo Sánchez Sorondo, eds. pp. 391–394. Cham, Switzerland: Springer. The quote is from p. 393.

23 Murphy, Earl. 1967. *Governing Nature*. Chicago: Quadrangle Books.

24 Grasso, Martin, and Richard Heede. 2023. "Time to Pay the Piper: Fossil Fuel Companies' Reparations for Climate Damages." *OneEarth* doi 10.1016/j.oneear.2023.04.012.

25 Smith, Hayley. 2023. "Most of World Is Feeling Climate Change." *Los Angeles Times*, Aug. 3, A1, A4.

26 Adams-Fuller, Terri. 2023. "Dangerous Discomfort." *Scientific American*, July/August, 64–69.

27 Ballester, Joan, Marcos Quijal-Zamorano, Raúl Fernando Méndez Turrubiates, Ferran Pegenaute, François R. Herrmann, Jean Marie Robine, Xavier Basagaña, Cathryn Tonne, Josep M. Antó, and Hicham Achebak. 2023. "Heat-Related Mortality in Europe during the Summer of 2022." *Nature Medicine* 29:1857–1866.

28 Jørgensen, Lisa B., et al. 2022. "Extreme Escalation of Heat Failure Rate in Ectotherms with Global Warming." *Nature* 611:93–98.

4

PEOPLES IN PERIL

"Injustice anywhere is a threat to justice everywhere. We are caught in an inescapable network of mutuality, tied in a single garment of destiny. Whatever affects one directly, affects all indirectly." (Martin Luther King)

In Season 2, Episode 16 of "Star Trek: The Next Generation," the crew of the Starship Enterprise first encountered the "Borg," a cybernetic society that absorbs other individuals and societies into "the collective," homogenizing its members into a monolithic entity that then moves on to absorb the next society. If a society (or individual) does not comply, they are destroyed ("resistance is futile").

While this is a creation of science fiction, it aptly describes Western society in its relationship in colonizing small traditional societies. The colonization process includes initial contact, an assessment of whether the traditional society has something of value (land, timber, gold, etc.), invasion (often by companies backed by military force, sometimes using genocide), assimilation of the survivors (often assisted by missionaries), disease, poverty, and then generally servitude. Although some individuals of the small societies may survive, their societies do not, instead becoming "Westernized." While colonization has occurred in many places at many times, Western society's version is particularly aggressive and unfortunately quite effective (to the detriment of others).

The alteration or destruction of other societies impacts both the natural and cultural environment in a variety of ways. The destruction of one group will affect the cultural landscapes of surrounding groups by altering their economies, shifting social, political, and military alliances, and perhaps even leading to warfare, all of which can then impact the natural environment. Coupled with this is the loss of knowledge of the destroyed group.

DOI: 10.4324/9781003560326-4

While many people may be vaguely aware of this, it may seem to be of little concern. It may be that the demise of a small society in the depths of the Amazonian rainforest will not directly impact gasoline prices, but their loss may ripple across the planet. For example, nearly half of the medicines used in the modern Western world originated from plants used by and knowledge recorded from medical practitioners in small traditional societies (think of this next time you take an aspirin).[1] The loss of those practitioners means the loss of their knowledge and possibly the next "miracle drug" that might end up saving you or a loved one's life. Also, from a moral and ethical standpoint, ignoring injustice makes one culpable.

Anthropology teaches us that all societies are valid, have the right to exist, and that it is wrong to judge them and immoral to destroy them; a principle called cultural relativism (there are exceptions; it is OK to harshly judge Nazi Germany and similar behavioral disasters). This is the same principle that makes it immoral to murder a person since they too have the right to exist.

An associated issue is ethnocentrism, the view that one's society or group is better than others. All societies have this view and sometimes even view non-members as not even being human. Americans are no exception, and this belief system is twisted to justify the denigration and systematic discrimination of minority and ethnic groups in the United States.

Early in the history of Anthropology, a theory was proposed about how societies evolved, beginning as "savages" (e.g., hunter-gatherers), advancing to "barbarians" (e.g., simple farmers), and finally to "civilized" (complex societies with cities and writing).[2] While this theory was rejected by anthropologists, the derogative descriptive terms live on and continue to be used to rationalize the mistreatment of other societies. The use of the term "primitive" has the same detrimental impact. For example, early Euroamerican settlers in North America considered Native Americans to be "savages" who were impeding the expansion of white "civilization." As a result, many Native people were killed, moved, or incarcerated either directly by the government or with their approval. Today, countries attempting to become "develop" treat their indigenous people in the same way.

During the 1950s, nations were divided into political "worlds." The First World was used to describe countries that were aligned with the US and the West; the Second World was the communist states; and the Third World was the unaligned states. With the end of the Cold War, the terms First and Second Worlds have been dropped from usage, but the term "Third World" is still in common usage and now refers to developing countries.

Some anthropologists use the term "Fourth World" to refer to extant indigenous societies living as dependents within contemporary countries.[3] There are some 300 million fourth-world Indigenous people in some 5,000 groups, speaking 4,000 languages, and living in 90 countries. Well known examples include nearly a thousand different Native American societies in the United States and Canada and about 500 individual Indigenous Australian societies. A Fifth World consists of past societies known mostly through archaeology.[4]

Colonization

Colonization is the process of a species occupying a new area or region. The initial human colonization of the planet began perhaps several million years ago when an assortment of human species (e.g., *Homo erectus*, archaic *Homo sapiens*, Neanderthals, and Denisovans) spread across most of the Old World (Africa, Europe, and Asia), which was then unoccupied by people. Modern humans (*Homo sapiens sapiens*) emerged from Africa about 70,000 years ago and recolonized the Old World, replacing (killing or absorbing) most of the other humans (those that survived the effects of the eruption of the Toba supervolcano[5] in Indonesia) to become the dominant and ultimately the only human species.

Since then, groups of people have colonized other groups of people many times over. Beginning about 10,000 years ago, farmers began to colonize the lands occupied by hunter-gatherers, pushing them out of "good farming lands" and into marginal areas. In Europe, this process took several thousand years, resulting not only in the deforestation of much of Europe but also in the destruction of the resident hunter-gatherer groups. Today, very few hunter-gatherer groups have survived the onslaught of the farmers, and they persist only in areas where farming is impractical, such as arid deserts and the Arctic.

Farmers can also be the victims of colonization. When Europeans began to colonize the Americas (the New World) in the early 1500s, the resident Native peoples were mostly farmers, some with highly complex societies with cities larger than any in Europe at that time. The Spanish colonized much of Central and South America, but their primary goal was to extract wealth from the Native peoples and to convert the people to Christianity rather than to occupy their land.

First, the Spanish "pacified" the Native populations through ruthless and brutal campaigns of violence, slavery, and extermination, leaving no doubt as to who was in charge. European diseases further reduced Native numbers by as much as 95 percent.[6] Once the Native peoples were pacified, the Spanish established a feudal system, giving huge tracts of land to certain Spanish nobles who then extracted labor and wealth from the resident Native people, using military force if necessary. This system was later modified with a scaled-down version with smaller tracks of land called haciendas. This system was finally broken up after the Mexican Revolution of 1910.

The English (British) and later the Americans approached the colonization of North America quite differently. They did not want to exploit Native labor or resources; they wanted the land to provide living space for their landless and/or unemployed citizens (plus criminals exiled to the colonies). The colonists considered the Native peoples an obstacle, to be pushed out or killed to make way for still more colonists. This pattern of physical and cultural genocide continued after the newly formed US took over relations with the Native Americans. The British followed this same basic policy in their colonization of Australia but used the Spanish model in India.

Genocide and Ethnocide

Genocide is the killing, or attempted killing, of an entire ethnic group, religious group, or nation. Like colonization, genocide has been a part of the human experience for a long time. Modern humans undoubtedly exterminated their human cousins as they colonized the Old World, and there are other examples from archaeology. In more recent times, examples of physical genocide include the Armenians during World War I, the Jews in World War II, the Tutsi in Rwanda in 1994, the Bosnians in 1995, and the ongoing genocides in China (the Uyghur), Myanmar (the Rohingya), and Ukraine (by the Russians).

Ethnocide, the destruction of a cultural system, also has a long history. When entire groups of people are killed (genocide), their culture typically dies with them. But in ethnocide, the people physically survive but are assimilated into the dominant (colonizing) society with the effect that their culture, language, and belief systems are destroyed. A loss to us all.

There are many contemporary examples of ongoing genocide and/or ethnocide. For example, in the Amazon, lumber and mining companies consider the Indigenous people to be in the way of "development" (read exploitation) of the resources of the forest. Native men are killed or enslaved, women are forced into prostitution, Native villages are literally bombed using drones, and activists are murdered. All this has the tacit approval of the national governments and local police (if any, as much of the Amazon is the "Wild West") since they make money from the companies. A horrible situation.

Major Issues for Traditional Societies

All traditional societies reside within formal countries with national governments. As such, they have all been colonized by the dominant society to some degree. Some of these Fourth World societies are now faring relatively well (e.g., those in North America but only after a terrible history), while others (e.g., in Brazil) remain under severe and violent pressure. Nonetheless, all face many of the same basic issues.

Sovereignty and Decolonization

Prior to being contacted by colonizers, traditional societies were sovereign and independent groups, making their own decisions and managing their own affairs. After colonization, the ability of these groups to make decisions was negated, and they became dependents of the dominant societies. This lower status was, and continues to be, reinforced by efforts to extinguish traditional practices, religions, languages, and social structure. Traditional peoples lost control of their land, lives, livelihoods, and identities.

Many Native people have been placed on reservations, reserves, camps, and the like. Often accompanying this are restrictions on movement, religion, and languages, plus dependence for food and health care; in essence, the complete subjugation of Native culture. In some cases, Native people are little better than slaves working for companies. In all cases, Native people must subvert themselves to the dominant society to get anything. Anyone resisting is generally classified as "hostile," "rebel," or most recently "terrorist" and is dealt with by force.

In most cases, traditional societies strive to regain sovereignty; to reassert control over their own lives, and to regain political and economic independence.[7] The re-establishment of sovereignty is at the heart of decolonization. However, as one might predict, decolonization is a daunting challenge since colonization is still ongoing in many places. Dominant governments will rarely return confiscated lands and property or relinquish power to anyone, much less Native people. However, in a few places, decolonization efforts have had some success.

For example, in the United States, Canada, and Australia, considerable power has been returned to Native groups, although they remain dependent nations. In 1999, Canada established a new province called Nunavut, governed by the Inuit, although still under the Canadian government. Perhaps more remarkable is in Greenland, where the Inuit Ataqatigiit party won a majority in the Greenlandic legislature in 2021 with the goal of gaining independence from Denmark.

Land Claims

A major issue for traditional peoples has always been the loss of land (and water and mineral rights), the confiscation of which is an integral part of colonization. Native lands are frequently altered by their new "owners" to the detriment of biodiversity and other aspects of the natural environment. The loss of the land also results in the inability of traditional groups to continue their economic practices, stressing both the traditional system and that of the dominant society.

The confiscation of land by dominant groups continues today. For example, in 2019, the government of India mandated the removal of some eight million Indigenous people (collectively called the Adivasi) from their forest homes, ostensibly to "preserve" the forest. However, the Adivasi are much better forest managers than the corporate entities that are replacing them, and this will likely degrade, not preserve, the forest.[8]

Associated with land rights is access to traditional natural resources that have been utilized for millennia. Access to hunting, fishing, and plant-gathering areas is an important issue. In the US, many treaties were signed that guaranteed the rights of Native groups for fishing, hunting, and gathering, but these rights were largely ignored. Some of these rights are now being reclaimed. For example, Native American fishing rights in the state of Washington were reaffirmed in 1974 and upheld by the US Supreme Court in 1979.

There has been some progress in land rights. The United States established a land claims commission to deal with such claims. For example, in 1971, a comprehensive settlement was reached with Native groups in Alaska, and 44,000,000 acres were returned to Native groups, along with a cash payment of $962 million and a royalty of $500 million on mineral rights. The Canadian government has also signed several land claim settlements, as has the Australian government.

Violence

Violence against indigenous people has a very long and sad history. Violence, including rape, murder, and genocide, is a tool of colonization and continues to be a major issue among traditional societies. Such violence disrupts a society and its ability to adapt and persist. Another tool of colonization is assimilation, and the forced assimilation of children, often accompanied by violence, was and remains a common practice. In many instances, children were forcibly removed to boarding schools where they were taught the language and customs of the dominant society in an effort to eradicate Native traditions and practices. Such efforts not only impacted the children but also their families and subsequent generations.

Another pattern is the violent rape and murder of women. In some cases, girls and women are forced into prostitution and/or slavery. In other cases, they simply disappear. In the US and Canada, there is an epidemic of violence against Native women, but this is often ignored by the dominant governments.[9] For example, the murder rate among Native women in Canada is as much as six times higher than for other women, and the issue is rooted in the apparent unwillingness and inability of the state to deal with the problem, seemingly due to general colonial attitudes against Native people. As one could imagine, this is a horrible fate for the individual women and very disruptive to their society.

Loss of Language

There are currently some 5,000 small-scale societies on Earth, mostly with distinctive cultures, religions, and languages. Language is one of the hallmarks of a society, and it encodes a great deal of information, such as history, traditional knowledge, philosophy, and identity. The suppression of Native languages is one of the goals of colonial assimilation, beginning with children forcibly removed to boarding schools. Other causes of language loss include the death of speakers (e.g., from disease and violence) and the necessity of using the dominant language in dealing with the authorities.

The preservation of indigenous languages is a worldwide issue. Many Indigenous languages are now extinct, while many others are critically endangered, with only a few speakers remaining. Numerous efforts are underway to record and catalog Native languages and to establish classes to teach young people their own language.[10]

Loss of Knowledge

Traditional societies possess vast amounts of knowledge. Some of this knowledge is cultural, related to religion, social structure, history, art, morality and ethics, and the like. Other major categories of knowledge include botany, zoology, astronomy (e.g., the Maya calendar is more precise than the Western one), medicines, farming systems, and edible foods (for example, of some 4,000 varieties of potato, only about a half dozen are in the Western grocery stores). The storehouse of Indigenous knowledge is vast and could be extremely useful to the rest of the people on the planet. Some of the traditional agricultural methods could be adapted and used by Western farmers (see Chapter 7).

The loss of this knowledge, often encoded in language, is a major issue. One problem is that we do not know what they know and so do not understand what is being lost. It is like being given a present wrapped in a box and then tossing it into the trash. What was in it? We will never know. But let's not throw the next box away.

Theft of Intellectual Property

Traditional knowledge, including medicines and practical applications useful for agriculture and other fields, is vast. In many cases, this knowledge is being used by Western companies without compensation to the knowledge holder, a sort of copyright or patent infringement. This could be seen as an extension of Western colonial practices and exploitation of Native peoples. Ways to deal with this issue, such as extending "copyright laws" or royalty agreements to include unwritten knowledge, have been proposed.[11] This issue requires further work for an equitable solution.

Loss of Religion

The suppression and replacement of Indigenous religions with Western ones (mostly Christianity) is both a tool and a consequence of colonization. Religion is a fundamental aspect of any society and is not compatible with assimilation; therefore, the colonizers must replace it. Considerable knowledge is encoded in religion (including Christianity) as is lesson in morality and ethics.

The suppression of Native religions is ongoing, reinforced by missionaries who establish missions within Indigenous communities to try and convert them. The Native people have little power to prevent this. In their zeal to convert people, the missionaries also impact other aspects of their society, imposing new values and restrictions to comport with Western practices. In 1994, the US made it illegal for the federal government, but not missionaries, to interfere with Native religions.

Climate Change

Climate change affects all people, everywhere but perhaps with a disproportionate impact on Indigenous people who are dependent on the actions of the dominant society and others over which they have no control. Forests are disappearing, drought is becoming more common, food insecurity is increasing, sea levels are rising and drowning small island nations, heat is increasing, and resources used for millennia are becoming more difficult to find. Traditional farming is becoming more difficult as water supplies dwindle.

In the Arctic, the reduction in sea ice has resulted in changing the types and numbers of animals available, such as the decline in seal populations that have impacted the food sources for both people and polar bears. Rising sea levels are causing increased coastal erosion and inundation, threatening the destruction of traditional coastal villages and cemeteries.

The reduction of sea ice in the Arctic foreshadows other consequences. Oil and gas are easier to find and extract, and there is now an intense effort to find and exploit these resources by the US, Canada, Russia, and Denmark. In addition, the long sought-after "Northwest Passage" may soon be usable for at least part of the year, increasing commerce. These developments will significantly impact Arctic peoples and biosystems.[12]

Saving Lives and Societies

What can be done to protect traditional peoples and their environments? Probably the primary issue is the involvement of national and local governments in the problem, either through bigoted indifference or as coconspirators with commercial interests. In many instances, Indigenous people are not even citizens of the country they live in, making the denial of their rights easier to justify internally. Native people were not made citizens of the US until 1923 and not in Australia until 1967.

Like so many issues regarding the environment, governments will actively work against environmental protection, deny and ignore the problem, or simply fail to act even if it is clear that they should. Thus, it is necessary to motivate government officials to do the right thing through public awareness, protests, and/or voting for more responsible people.

A number of international organizations exist to educate the public and provide support for Indigenous people. One such organization is *Cultural Survival*, dedicated to the human rights of Indigenous people. Other organizations purchase lands for Native groups with the added benefit that the land will not be "developed." There are also activists on the ground educating Indigenous people and providing them the tools to resist. But this is a very dangerous profession, as many are murdered, likely by people tied to commercial companies.

In some instances, Indigenous peoples are acting in their own interests. For example, in Australia, some Indigenous groups are leaving the confines of the stations (cf. ranches or towns) where they live, work, and interact with the government. These groups are going back to ancestral lands, adopting their traditional lifestyle, economies, and religions.[13] Indigenous Australians lived in these settings for at least 60,000 years without Western help, and they can do so again.

Another such example comes from the Canadian Arctic. There, some Inuit families or groups have returned to a traditional mobile hunting lifestyle using sleds and dogs rather than snowmobiles that require an extensive logistical network for operation and maintenance. This alone severs most of the connections of those people with the Canadian government and other Western goods. This traditional system worked for more than a thousand years. It is a hard life, but it is theirs.

Notes

1 Salmerón-Manzano, Esther, Jose Antonio Garrido-Cardenas, and Francisco Manzano-Agugliaro. 2020. "Worldwide Research Trends on Medicinal Plants." *International Journal of Environmental Research and Public Health* 17(10):3376; Wiley, Andrea S., and John Scott Allen. 2020. *Medical Anthropology: A Biocultural Approach* (4th ed.). Oxford: Oxford University Press; U.S. Forest Service, "Medicinal Botany," USFS website, Medicinal Botany (usda.gov).
2 See the extended discussion in Harris, Marvin. 1968. *The Rise of Anthropological Theory: A History of Theories of Culture.* Chicago: University of Chicago Press.
3 Neely, Sharlotte, and Douglas W. Hume (eds.). 2024. *Native Nations: The Survival of Fourth World Peoples* (4th ed.). Vernon, British Columbia: J Charlton Publishing, Ltd.
4 Sutton, Mark Q. 2017. "Voices from the Past: Conceptualizing a 'Fifth World.'" *Journal of Anthropology and Archaeology* 5(1):17–19.
5 Williams, Martin A. J., Stanley H. Ambrose, Sander van der Kaars, Carsten Ruehlemann, Umesh Chattopadhyaya, Jagannath Pal, and Parth R. Chauhan. 2009. "Environmental Impact of the 73 ka Toba Super-Eruption in South Asia." *Palaeogeography, Palaeoclimatology, Palaeoecology* 284(3–4):295–314.
6 Cook, Noble David. 1998. *Born to Die: Disease and New World Conquest, 1492–1650.* Cambridge: Cambridge University Press.
7 Lerma, Michael. 2014. *Indigenous Sovereignty in the 21st Century: Knowledge for the Indigenous Spring.* Gainesville: Florida Academic Press.
8 Neely, Sharlotte. 2024. "What It Means to Be Indigenous." In *Native Nations: The Survival of Fourth World Peoples* (4th ed.), Sharlotte Neely and Douglas W. Hume (eds.), pp. 1–6. Vernon, British Columbia: J Charlton Publishing, Ltd.
9 Ficklin, Erica, Melissa Tehee, Racheal M. Killgore, Devon Isaacs, Sallie Mack, and Tammie Ellington. 2022. "Fighting for Our Sisters: Community Advocacy and Action for Missing and Murdered Indigenous Women and Girls." *Journal of Social Issues* 78(1):53–78.
10 McDermott, Brian. 2014. "Language Healers: Revitalizing Languages, Reclaiming Identities." *Cultural Survival Quarterly* 38(1):18–19.
11 Laird, Sarah A. (ed.). 2002. *Biodiversity and Traditional Knowledge: Equitable Partnerships in Practice.* London: Earthscan.
12 Anderson, Alun M. 2009. *After the Ice: Life, Death, and Geopolitics in the New Arctic.* New York: Smithsonian Books.
13 Peterson, Nicolas, and Fred Myers (eds.). 2016. *Experiments in Self-Determination: Histories of the Outstation Movement in Australia.* Canberra, Australia: Australian National University Press.

5

THE AIR WE BREATHE

Landing in LA in the 1970s, there was so much smog that it looked like we were descending into a mud puddle, MQS (co-author Sutton) personal observation.

The atmosphere is the covering of gases that envelop the Earth. It is largely held in place by gravity (some other planets and moons also have atmospheres). On Earth, the atmospheric gases consist of about 78 percent nitrogen, 21 percent oxygen, 1 percent argon, 0.04 percent carbon dioxide, and trace amounts of hydrogen, helium, and other gases. The atmosphere also contains a great deal of water vapor plus aerosols, small particles such as dust, pollen, soot, smoke, other pollutants, and bioaerosols such as microbes. The atmosphere keeps the planet warm enough to support life, and the ozone layer protects life from too much ultraviolet radiation.[1]

The atmosphere of Earth is composed of five major layers, each with different properties. The troposphere is the lowest layer and varies in height between 17 km at the equator and 7.0 km at the poles and contains the bulk (ca. 75 percent) of the mass of the atmosphere. Weather generally occurs in the troposphere, and this layer is very dynamic, with air moving around due to temperature differences and storms.

The stratosphere extends between 15 and 35 km from the top of the troposphere and contains the ozone layer. The mesosphere ranges from 50 to 85 km. The thermosphere exists between about 85 and 690 km and contains the ionosphere that will reflect radio waves. Finally, the exosphere begins at about 690 km and extends to roughly 10,000 km.

The atmosphere weighs about 5.5 quadrillion tons (5.5 followed by 15 zeros) and exerts a great deal of pressure, increasing as one moves lower in the atmosphere. As pressure increases, heat increases, so it is always warmer lower than

DOI: 10.4324/9781003560326-5

higher. Also, as one moves higher, the density of the atmosphere decreases, and the thinner air makes breathing more difficult. Anyone who has visited Machu Picchu in highland Peru knows it takes a few days to adjust to the lower air pressure.

A Lot of Bad Air

Pollution is the presence of something that is harmful. In the case of air, harmful materials originate from two major sources: nature and humans. Natural air pollution includes dust (soil particles, some of which are toxic), smoke from natural fires, ash from volcanoes, and the like. Human-caused air pollution is primarily from the manufacture and use of industrial products. For example, a car requires the manufacture of metals and plastics, assembly, transport, sales, maintenance and parts, fuel, and finally disposal or recycling. Each of these steps results in the production of pollutants that are then released into the air and water. Multiply this by all the other manufactured items, and one can see how massive the pollution problem is.

Among the major sources of ambient (outdoor) air pollution is the burning of fossil fuels. Car and truck exhaust and the burning of coal to produce electricity are common sources (Figure 5.1). These generate not only greenhouse gases but also nitrogen and sulfur. They also produce particulates, which are intensely unhealthy for the respiratory system. Exceedingly small particles are among the deadliest; diesel engines are continually being improved to reduce their production of these.[2]

MEHR NEWSAGENCY
Photo: Abd Fathi

FIGURE 5.1   Air pollution in the city of Karaj, Iran, in 2021. Photo by Mehr News Agency (Wikimedia Commons, Public Domain).

The use of natural gas in electrical generation plants is far less polluting and emits about half the amount of GHGs of coal.

The major forms of air pollution include particulates (e.g., soot from diesels), carbon monoxide, ozone, nitrogen dioxide, and sulfur dioxide. One of the biproducts of burning coal is the emission of sulfur dioxide into the air, which then combines with atmospheric water to produce an acid that then falls back as "acid rain" that then contaminates soil and water supplies, adversely impacting plants and animals. Another major emitter of sulfur dioxide is beef production. A further class of ambient pollutants in urban areas is drugs such as cocaine, heroin, cannabinoids, and amphetamines.[3] In addition to ambient air pollution, indoor household air pollution is also a major issue. In the US, household air contains formaldehyde, radon, microplastics, bacteria, hair, pollen, fragmented dead insects, dust (including dead skin cells, bits of hair, dust mites, and dust mite feces), and perhaps tobacco smoke and resins. Elsewhere, around 2.4 billion people cook inside using kerosene, wood, animal dung, or coal in open fires or stoves, all of which are polluting. Gas stoves are much better but still cause health problems in confined, poorly ventilated places. According to the World Health Organization in 2023, household air pollution "was responsible for an estimated 3.2 million deaths per year in 2020, including over 237 000 deaths of children under the age of 5" by loss of lung function.

Combined, ambient and household air pollution result in 6.7 million premature deaths annually, and smoking kills another six or seven million.[4] Air pollution is especially bad in China. Beijing has had to shut down several times in recent years due to the heavy pollution. Shanghai and other large cities are similarly troubled. Acid rain is killing vegetation around them. Recent efforts have decreased air pollution, but there is a long road ahead. A similar situation exists in Europe. India also has problems, with Delhi possibly having the world's most polluted air now. Mexico City, like China's cities, has been improving but is still substantially polluted.

The air quality in the US was also very bad, but the passage of the Clean Air Act in 1963 (as amended) has resulted in significant improvements. Here in southern California, we can see the mountains again—they were usually invisible from the cities back in the 1960s. We remember when new students at our universities were surprised to learn, when a winter storm cleared the skies, that there were mountains above them. However, there is a recent effort by conservatives in Congress to weaken the law to allow for more industrial pollution.

The southern San Joaquin Valley of California has some of the worst air in the US. This is due to its geographic location at the end of a valley, exhaust from many cars and trucks, emissions from local oil production, local farms creating dust and spraying pesticides and herbicides, and the accumulation of pollutants that flow into the area from cities to the north, such as San Francisco. Many thousands of people become ill each year as the defoliants are sprayed on the cotton fields.

Ambient air pollution kills half a million children in Africa every year.[5] The worst problem is natural dust blowing in from the desert. However, industrial pollution is increasing and dangerous. A green wall of trees south of the Sahara is slowly growing, but funds are limited and the environment is harsh, so the dust will not go away soon.

Another enormous problem in Africa, and also in Asia and parts of Latin America, is cooking on open, smoky fires or poorly designed stoves. The fumes are deadly. Even modern gas stoves are polluting enough to be facing legal restrictions in California. Consider how much worse it is to cook over an old-time camp stove or kerosene stove. At least open fires usually go with well-ventilated houses, but, if not, the results can be deadly. Carbon monoxide is the most dangerous offender, but particulates cause damage to lungs and respiratory systems. Supplying better stoves has become a high priority in rural Africa and Asia. Electric stoves are best (currently), but even the fuel-efficient bucket stoves of East Asia are a huge improvement. These are simple: line a bucket with a layer of fireclay or the like, a couple of inches thick, and cut two holes in the bucket for fuel and exhaust. The top becomes a rest for a cooking pot. ENA has seen more than one four-course dinner cooked on such a stove in Hong Kong. Better are the efficient methane stoves, running on methane generated by agricultural and human wastes, rapidly increasing in popularity. These will be good stopgaps while we wait for more hi-tech and energy-efficient stoves to reach the rural world.

One pollutant not often considered is the enormous amount of nitrogen that comes from burning fossil fuels. Car exhaust fertilizes the world. It has led to dramatic changes in California's vegetation. Fires lead to replacement of native plants that fix their own nitrogen or manage on short supplies by nitrogen-craving nonnative weeds—grasses and mustards especially—that crowd out the natives and burn more readily.

Nitrogen from air and fertilizer also gets into the waters of the world, causing eutrophication and algae blooms. The world is getting cheap crop boosts from fertilizer, but the damage to wild ecosystems is serious and potentially devastating.

Among odd byways of pollution are the countless gigantic freighters, container carriers, tankers, and cruise ships. Many of these are registered under "flags of convenience," that is, in countries that do not have meaningful regulation to stop pollution and dumping. The European Union fleet is almost all so registered, with the Comoros, Palau, Bahamas, Panama, and Liberia being especially popular. Panama and Liberia have had this reputation for decades. Most of the old-time Hong Kong freighters used them. These ships, often huge, are vast moving pollution-mills.[6] One container ship can produce as much pollution as several thousand cars.

Recently, wildfires in places such as the western US, Canada, and Russia are having a dramatic effect on air quality worldwide. These fires are the result of some poor management (suppression of natural fires) and the associated accumulation of duff (dead materials that provide fuel to a fire) in the forests, the drying of the vegetation from drought, and the death of many trees from insect infestation enabled by drought. The smoke and ash from these fires had created a major health hazard. In addition, vast amounts of previously stored carbon are being released back into the atmosphere. The chemicals and particulates in smoke are exceedingly bad for human health.

Making matters worse, the shrublands and grasslands of the world have been invaded by highly flammable grasses.[7] This was a major cause of the wildfire that destroyed Lahaina, Maui, Hawaii, in 2023. Cheatgrasses (*Bromus* spp.) and *Ventenata dubia* in the western US, buffelgrass in the American Southwest and Mexico, molasses grass and *Imperata* in the tropics, and other such grasses are turning much of the world into regions occupied by single species of flammable grasses. This is the result of two things: the human introduction of these grasses and suppression of the natural fire regime of frequent small fires. Fire suppression leads to fuel buildup and enormous unmanageable fires when the inevitable finally happens, as is the case in forests.

Even light is an atmospheric pollutant now. Cities illuminate the night so brilliantly that birds are lost and fly into buildings, night-flying insects cannot behave normally and die out, and other natural cycles and patterns are massively disrupted, to say nothing of astronomers' work. Lights with narrow bandwidths would solve much of this. In the end, the nightless city is unsustainable; it simply draws too much energy.[8]

What to Do?

The atmosphere suffers from two major traumas: carbon and pollution. Carbon is harmless in itself; its problem is causing global climate change. Pollution is that which is actually dangerous to human health and other lives.

Reducing carbon emissions through various means would help mitigate the climate crisis. In brief, needed action on carbon includes shifting energy production from fossil fuels to green renewables (wind, solar, nuclear, and others) and the planting of a billion or so trees and other vegetation to store carbon from the air. Many more specific ideas are discussed in Chapter 11.

Regarding pollution, shifting energy production to green renewables would remove the pollutants from vehicles and coal and natural gas-powered generation plants, and that alone would considerably decrease air pollution. Stabilizing soils would help reduce dust. Moving away from so many airborne agricultural poisons would also help. Basically, the bottom line is minimizing the stuff we put in the air. That requires controlling factory exhaust, vehicle exhaust, and other industrial processes, as well as large-scale farming. The technology is largely developed and in place in some areas. As usual, the problem is getting politicians to pass necessary legislation to mandate such measures.

Notes

1 Sanchez-Lavega, Agustin. 2010. *An Introduction to Planetary Atmospheres*. New York: Taylor & Francis.
2 Smil, Vaclav. 2017. *Energy and Civilization*. Cambridge: MIT Press.

3 Viana, Mar, Xavier Querol, Andrés Alastuey, Cristina Postigo, MJ López de Alda, Damià Barceló, and Begoña Artíñano. 2010. "Drugs of Abuse in Airborne Particulates in Urban Environments." *Environment International* 36(6):527–534.

4 World Health Organization. 2023; *Science*, news item, 2018.

5 Heft-Neal, Sam, Jennifer Burney, Eran Bendavid, and Marshall Burke. 2018. "Robust Relationship between Air Quality and Infant Mortality in Africa." *Nature* 559:254–258.

6 Schiermeier, Quirin. 2021. "Boom in Ships with 'Fake' Flags that Trash the Environment." *Nature* 594:13.

7 Cornwall, Warren. 2022. "Fiery Invasion." *Science* 377:568–571.

8 Morgan-Taylor, Martin. 2023. "Regulating Light Pollution: More than Just the Night Sky." *Science* 380:1118–1120.

6

THE WATER WE DEPEND ON

Whiskey's for drinking, water's for fighting (attributed to Mark Twain, though not found in his writings)

All known life depends on water. While it seems that there is water almost everywhere, the distribution of water may be surprising. Most (97 percent) of water is in the oceans, with only about three percent being fresh (non-salty). Of the freshwater, most (69 percent) is held in glaciers, followed by groundwater (21 percent), with remarkably little (1.2 percent, less than one-half of one percent of all water on Earth) as surface water, permafrost, lakes, and rivers. About 20 percent of the surface water is in lakes, with Lake Baikal in Russia being the largest single source, with the African Great Lakes, the North American Great Lakes, and other lakes containing most of the rest. All this means that of the total amount of water on the planet, only a very, very small percentage is available for human use, and two-thirds of that is used by farmers.

Freshwater is replenished through precipitation, but if people use more water than is replenished, water becomes less available, not only to people but to all the other users of water, such as plants and animals. In addition, pollution will decrease the amount of usable water. People are all too prone to use rivers as sewers (Figure 6.1), and to dispose of chemicals in streams, lakes, and oceans. The bottom line is that there is precious little water for people to use.

People depend on water to drink, to clean, to dispose of waste, to keep their animals, for farming, and for wild plants and animals to endure. A lack of water in any of these areas spells death. Humans are about 60 percent water by volume, slightly less in women and more in men and children.[1] If a person becomes dehydrated and loses more than about 15 percent of their water, they will likely die. At least 10,000

DOI: 10.4324/9781003560326-6

FIGURE 6.1   Pollution in the Kathmandu River, Nepal, 2012 © Michel Royon (Wikimedia Commons, Public Domain).

people die from dehydration every year in the US, many more worldwide. It is estimated that about 25 percent of the world's population lacks access to safe drinking
water. Every year, some 800,000 people worldwide die from drinking unsafe water.
How much pollution in your drinking water would you tolerate?

Water is already in high demand. In 2023, some two billion people live in areas
where water is scarce. This number is rapidly rising, and demand is expected to
increase by 40 percent by 2050. Good reviews of water issues include *When the
Rivers Run Dry* by Fred Pearce and *The Three Ages of Water* by Peter Gleick.[2]
Water used by people comes from two basic sources: groundwater and surface
water such as rivers, streams, lakes, and reservoirs.

Tapping the Ground

Water from rain either evaporates, flows to the ocean, or soaks into the ground.
Water soaking into the ground will accumulate in underground basins, forming
aquifers. The water within an aquifer might overflow underground into other aquifers. The top of the aquifer, or water table, may be shallow or deep. If the water
table is very shallow, water may come to the surface in springs, artesian wells, or
flow into streams and rivers. The water in aquifers is mostly old, having filled the
basins long ago and is called fossil water. If the water removed does not exceed the
recharge from rainfall, the water table will remain stable, and the use of the aquifer
will be sustainable.

In the past, and in the present in many places, people generally accessed this underground water by digging wells into the water table and removing the water they needed by buckets or hand pumps. Using this technology, it was difficult to remove large quantities of water from wells, and aquifers were largely stable. However, with the introduction of powered mechanical pumps, far more water can be removed than can be recharged, resulting in a lowering of the water table and even the depletion of aquifers.

Thus, groundwater everywhere is disappearing fast; wells are dug sinking ever deeper, and the aquifers are being drained. In portions of California's Central Valley, so much groundwater has been pumped out to feed the farms that the surface of the ground has subsided some 30 feet. In China, almost a million square kilometers of land is subsiding from groundwater overdraft.[3] On the plus side, groundwater recharge basins, where excess water is stored so it can sink back into the ground, have been established in some places.

Pumping Fossils

Deserts currently occupy much of North Africa and the Arabian Peninsula. The climate of this region cycles through wet and dry periods about every 20,000 years. Until about 5,000 years ago, the region was in a wet cycle, with considerable rainfall flowing into rivers and standing lakes and filling the large aquifers of the region. Once the wet cycle ended, the aquifers were full and are still there today.[4]

The governments of Libya, Egypt, and Saudi Arabia have discovered these ancient aquifers and are in the process of building massive projects to extract this water and pump it to coastal cities and to new and extensive farms. While this is great in the short term, it will rapidly deplete even the largest aquifers, with no hope of recharge for at least another 5,000 years.

Worse problems are faced by some other countries. Look at Iran, basically a large desert with extremely limited rainfall outside the Caspian Sea coast and its mountain wall. With a dense, young population of over 80 million and a diverse and flourishing agriculture, Iran is hardly able to live without water. Yet climate change is predicted to drop rainfall this century by 35 percent, as much as 80 percent in some places. The groundwater is poorly regulated, and Tehran alone has about 50,000 wells, many of which are illegal. About 80 percent of water used for agriculture comes from groundwater, which is being rapidly depleted, mostly fossil water that cannot be recharged. Iran's huge Lake Urmia is drying up and will soon be gone, with devastating effects on wildlife as well as people. The groundwater problem is so serious that in some regions, water must be trucked into villages and towns.

Withdrawals of groundwater in Iran went from 66 million cubic meters in 1965 to 133 billion in 2019. Not only exhaustion but also depletion of the freshwater down to saltwater level is getting critical. Only India, the US, Saudi Arabia, and China have more problems with groundwater. In Iran's central desert, people are now drilling down to 250-meter-deep groundwater. The entire Sistan province in eastern Iran may become uninhabitable since its groundwater is nearly exhausted and its surface water (the Helmand River) comes from Afghanistan and is already overexploited there. In most of Iran, it is now illegal to drill wells except for drinking water.[5]

What happens when these aquifers, with so many people depending on them, run dry? Hope for the best? Not a good policy.

Stressed Rivers

The other major source of water is rivers. In the river valleys of Mesopotamia some 10,000 years ago, early farmers began to develop irrigation systems and so became concerned about water supplies. If farmers upstream used all the water, the downstream farmers would have none. It was about this time that state-level societies began to develop. Among the many theories about how state-level societies developed was the "Hydraulic Theory" proposed by Karl Wittfogel,[6] in which the competition for water led to the development of complex management systems and the establishment of militaries to defend the water.

Of course, not all societies that used irrigation evolved state-level political structures, and some state-level societies developed without irrigation. Today, much river water is stored in reservoirs behind dams. Worldwide, there are perhaps 500,000 dams and reservoirs holding freshwater. The dams are generally constructed to control flooding, store water for homes and farms, produce hydroelectric power, and even provide recreation. Europe has over 1.2 million barriers—dams, weirs, river-blocking bridge foundations, and the like—on its waterways; density is highest in England, Germany, the Low Countries, and neighboring areas, but also high in Spain. This has virtually eliminated natural river flow in most of Europe, and with it the fish, marsh vegetation, and other biodiversity.

This is a hefty price to pay for damming a river. Natural ecozones are often destroyed, human communities may be flooded, archaeological sites inundated, navigation by ships may be restricted, and considerable water is lost through evaporation. Water flow in the river is disrupted, and the fertile silt it naturally carries will settle behind the dam rather than in the farm fields. This results in a loss of fertile sediments for downstream farms and an eventual silting in of the reservoir. It could get worse. The Grand Ethiopian Renaissance (or Millennial) Dam was constructed on the Blue Nile in Ethiopia and is now full and is curtailing the flow of the Nile as it passes through Sudan and Egypt. This loss of water has alarmed Egypt to the point of threatening war with Ethiopia.

Iraq also has issues. Turkey and Syria have built megadams on the Tigris and Euphrates Rivers, greatly reducing the flow of water into Iraq and thus impacting millions of Iraqis. Further, the Euphrates is drying up due to drought. To make matters worse, Iran takes 90 percent of the water of the Karun River, the only other major feeder river of Iraq. Rainfall is rapidly shrinking, and Iraq is warming even faster than the average for the region.

The Mosul Dam on the Tigris River is an issue. The dam, completed in 1987, now shows signs of instability, and in 2016, there was a warning to the people living in the floodplain below the dam about a possible collapse. If the dam were to collapse, between 500,000 and 1.5 million people would be at risk of drowning due to the subsequent flood.

The Aswan High Dam in Egypt is another concern. Built in the 1960s, the dam created Lake Nasser, a 310-mile-long reservoir that extends well into Sudan. The filling of the reservoir required the relocation of more than 100,000 people and flooded large tracts of farmlands and many archaeological sites. Some archaeological sites, including the famous temple of Abu Simbel, a major tourist attraction in Egypt, were moved out of the way. The dam controls the annual flooding, generates electricity, mitigates drought, has improved navigation on the river below the dam, and provides irrigation water.[7]

But here's the problem. For thousands of years, Egyptian farmers relied on the annual floods to deposit nutrient-rich sediments onto the fields, and this happened (mostly) without fail. Farming was very productive and was supplemented by the ubiquitous fish in the river and birds in the marshes. Because of the floods, villages were built in the desert just above the floodplain but close enough for farmers to tend their fields. People did not live on their farmland; that was for farming!

This all changed with the dam. There was no longer an annual flood, so settlements began to encroach onto the floodplain, reducing the available farmland (although this was partly mitigated by newly irrigated farmland). There was no longer the annual deposition of fertile sediments, so people had to begin to use chemical fertilizers, pesticides, and the like. While this did increase crop production, farmers now had to pay for the chemicals (the silt was free), and the river became polluted. This pollution impacted fish populations, and fishing in the river has greatly decreased (but increased in Lake Nasser). In addition, the river pollution has made the area where the Nile empties into the Mediterranean Sea a biological dead zone, destroying the fisheries that once thrived there. Adding insult to injury, the river water, the main source of drinking water for most Egyptians, is no longer fit to drink. And, of course, silt will eventually fill Lake Nasser. This is not sustainable in the long run. The ancient Egyptians had it right!

Consider also the Three Gorges Dam on the Yangtze River in China. Completed in 2006 (the hydroelectric part came online in 2012), the 1.4-mile-long dam created a huge reservoir stretching some 410 miles behind it. The dam greatly impacted the flora and fauna of the area, with many species going extinct while others became endangered. In addition, some 1.2 million people had to be moved from their traditional villages, leaving behind their sacred places and cemeteries, a major trauma to people who revere their ancestors. Further, increased deforestation in the area has increased erosion with the result that the reservoir is already beginning to silt in.

In addition to providing power, the dam was intended to protect the lower reaches of the Yangtze River valley from the flooding that has killed many millions of people through time, including about four million in the 1931 floods. This seems like a good idea, but if the dam fails, the resulting megaflood will endanger some 400 million people living below it, and there is already concern that the dam itself is deforming, increasing the risk of failure. The floods of the summer of 2024 fore-shadow the danger. Watch out![8]

Climate change and associated drought have decreased the flow of many rivers worldwide, exacerbated by the dams and reservoirs, leaving millions of people without adequate water. Examples of this include the Yellow River in China (so named from the color of the silt in the water) and the Euphrates River in Turkey, Syria, and Iraq. So much water is diverted from the Colorado River for homes and farms that it is effectively dry as it reaches Mexico, much to the displeasure of the Mexicans. The Jordan River, lifeblood for much of Israel and Syria and all of Jordan, is now not much more than a brook and a polluted one. The Amu Darya in Uzbekistan no longer reaches or feeds the Aral Sea, which has largely dried up, creating a desert from which toxic salts are blown by the ton onto nearby lands.[9] Half of China's 50,000 rivers and major streams are in danger of drying up.[10] A huge percentage of the world, including most of the US, and virtually all of Mexico, China, India, and the drier parts of Africa, is now stressed.

River deltas, in which a large percentage of the world's population resides, are also in extreme danger. This is caused by drying rivers, groundwater withdrawal, pollution, poor dike maintenance, poor erosion control, and similar problems. Climate change is adding to existing problems through the raising of sea levels and increasing intensity of storms.

The loss of freshwater biodiversity is another issue. The flooding of valleys behind dams has destroyed freshwater ecozones and is leading to a collapse of freshwater biodiversity. Only 0.16 percent of the world's rivers do not show this deterioration, but they are almost all in isolated Arctic regions.[11] Areas where dams and water systems have enabled consumers to have enough water despite short supply have done so at the expense of biodiversity; the dams and diversions dry up wetlands and distort the river ecology. As is expected, the Nile is a particularly scary situation, with 180 million people depending on its water. Most of its course is ecologically degraded, and from Cairo downstream, it is basically a sewer, as witnessed by MQS. Some short rivers now flow entirely within urbanized areas and have become urban drains (e.g., the Ogun River in Lagos, Nigeria).

Wild salmon are now a rapidly disappearing resource everywhere. The steel-head runs of southern California are down to a few fish; the only one south of Los Angeles is in San Mateo Creek, and it was down to a single breeding female fish in a recent drought.[12] Many, if not most, of the freshwater fish, amphibia, and shellfish of America are threatened or endangered. Caviar (the real kind, from sturgeons) will soon be a thing of the past, unless cultivated sturgeons succeed better than they are now doing; fishing for wild sturgeon is out of control, and sturgeons are succumbing to pollution and dams even where they are not fished. The healthy sturgeon populations in the world are in the major rivers of the Pacific Northwest, and even here they are declining fast.

One effect of channelizing and concrete-lining drainage ways is to prevent the water from sinking into the ground and recharging groundwater. A great deal of a stream's flow—in fact, all of it in much of California—is underground: the hyporheic flow. This zone extends well below and to the sides of the actual stream. It supports an enormous amount and variety of tiny creatures, from microbes to insects. Restoring creeks, including "daylighting" ones that had long been buried under concrete, would be a good step. This has been done in parts of Seattle and Los Angeles, with beautiful effect, as we have observed.

And It Gets Worse

Consider the water situation in the American Southwest, starting with Arizona. The largest metro area in Arizona is Phoenix and surrounding cities, all of which are expanding rapidly. However, the water locally available is exceedingly limited by climate and geography, and the area will be one of the most drastically drought-stricken areas of the world as climate change progresses. The groundwater is overdrawn and there is little recharge; the water now being pumped out entered the system more than 10,000 years ago. Phoenix has pumped out the last ten millennia of water in the last hundred years or so.[13] Phoenix is in big trouble but shows little sign of actually dealing with the problem. At least some surrounding cities have turned to desert landscaping instead of grass lawns.

The current situation in Arizona has an unpleasantly suggestive antecedent in the fall of the Native American Hohokam farmers that lived in the same area. The Hohokam and their neighbors constructed an incredible network of canals along the Gila and Salt Rivers. Some of these were as large and long as major contemporary irrigation canals. They fed an intensive agriculture based on corn, beans, squash, agaves, and many other crops. Sophisticated terracing and check-damming added to the water management picture. Yet, after devastating droughts in the 1200s, the Hohokam fields dried up or salted up, and the Hohokam disappeared.[14] The aptly named Salt River was not a good river to use for irrigation. Humans need a lot of salt, but most plants are killed by any significant amount of it in soil or water.

About 700 years ago, the Little Ice Age came, and the rivers refilled. The Pima arrived and made the land fertile and well-irrigated again. Unfortunately, the Spanish and then the Anglo-American settlers devastated this blissful scene by developing intensive irrigated agriculture, using increasing amounts of water. Finally, the Gila River went dry from Phoenix onward, leaving the Pima high and very, very dry, in violation of treaties as well as common decency. Ironically, the city of Phoenix takes its name from the ruins of the Hohokam city of Snaketown. The Americans planned a new city that would rise from the ashes of the ancient city just as the phoenix bird rose from its own ashes. The settlers were better prophets than they knew. The mythical phoenix is cyclically destroyed to rise again. We are about to witness the next fire.

Many places, including Phoenix, depend on water from the Colorado River. However, the entire western US is in a megadrought (with some improvement in 2023) with a significant decline in available water from the rivers and reservoirs. The flow of the Colorado has decreased about ten percent.[15] Recall that the draw of water from the Colorado River is such that it is dry before it reaches the Gulf of California in Mexico, in violation of treaties with Mexico. The Gila River, Arizona's major tributary of the Colorado, no longer comes even close to the Colorado except during abnormal flow. Indeed, most of Arizona's rivers are now dry washes for at least part of their length. Not long ago, the Santa Cruz River ran through Tucson, feeding mesquite thickets and the occasional cottonwood. No longer.

Las Vegas faces the same basic problem. The growing city is dependent on Lake Mead on the Colorado River for about 90 percent of its water. As the current drought is impacting the water level of Lake Mead, the water situation is precarious. The remaining water is taken from aquifers, but this is not an inexhaustible resource. The city has successfully implemented rules to lower water usage, but growth and overconsumption remain the problem.

The situation is similar in California. Ever-thirsty Los Angeles, along with most of southern California, imports water from several places. In 1913, construction was completed on the Los Angeles Aqueduct, a project that brought water to Los Angeles from Owens Lake in the southern Owens Valley, located some 150 miles to the north. This project has virtually destroyed Owens Lake, now mostly a dry playa whose lake sediments have been exposed to the wind, creating huge dust storms full of naturally deposited toxic materials. The courts have now required Los Angeles to leave enough water in the lake to at least keep the sediments moist so they will not blow away, but drought has made this a difficult process.[16]

In the 1960s, the state constructed the California Aqueduct to bring water from Northern California to Southern California. Colorado River water also feeds Southern California, but the recent drought has left those supplies under threat. Several coastal cities are building desalinization plants and attempting to better reclaim wastewater.

Another mess in Southern California is the Salton Sea, an inland lake located near the border with Mexico. Within the last 1,000 years or so, the Colorado River jumped its banks and flooded the large inland Coachella Valley, forming ancient Lake Cahuilla that extended from Palm Springs south into Mexico. The sudden appearance of the large natural lake full of fish and birds was a boon to the local Native Americans. Eventually, the river resumed its regular course and Lake Cahuilla dried up, only to form again, dry again, and form again. The last major stand of Lake Cahuilla ended about 300 years ago.[17]

In 1906, the US Army Corps of Engineers was in the process of channelizing the Colorado River but made a mistake, diverting the river into the Coachella Valley. The river flowed into the valley for two years before the levees could be repaired; thus, the Salton Sea was formed. It was a large freshwater lake teeming with fish and birds and became a favorite vacation spot for Los Angelenas, with resorts springing up along its western shore. Although the water began to evaporate, its level was stabilized by the addition of more water from the Colorado River and runoff from agricultural fields. However, this runoff brought with it pollutants including pesticides, herbicides, fertilizers, and other chemicals that eventually killed most of the life in the lake. The resorts were largely abandoned.

Since the Salton Sea is largely dead anyway, some want to let the it dry up so that the land could be farmed. Others want the lake preserved since it has become a part of the landscape. But the real issue is that if it was allowed to dry up, it would expose the sediments now toxic from farm chemicals to the wind (recall Owens Lake) and create a huge problem.[18] So, the state is forced to take considerable quantities of water from the Colorado River just to maintain the level of the lake and keep the toxins at bay.

And Even Worse Still

Pollution is also a serious problem in groundwater and rivers. For example, in Bangladesh and parts of Vietnam, so much groundwater is being pumped out that wells have to be dug deeper and deeper. But as they get deeper, the water is increasingly contaminated with natural arsenic found in lower sediments, making the water deadly.[19] Everywhere, though, agricultural and industrial wastes, including extremely toxic ones, are percolating into groundwater. About one-third of the Yellow River in China is so polluted from factory discharges and sewage that the water is not suitable for drinking, fish farms, industrial use, or even agriculture. The Nile River has similar problems. Not good.

In the US, the Mississippi River is polluted with chemical runoff from farms, primarily fertilizer but including pesticides and herbicides. The pollution increases as one moves downriver as more and more farm waste drains into the river. By the time it reaches the Gulf of Mexico, the water is so polluted that it has created a huge biological dead zone.

Lead poisoning from water pipes and other sources has been affecting people for millennia. The ancient Romans constructed systems of indoor plumbing that included piping in water through lead pipes. Unfortunately, the lead from the pipes contaminated the water, slowly poisoning the citizens of Rome. Some historians believe that this contamination may have hastened the fall of the empire, although others disagree.[20] In addition, the manufacture of lead pipes released lead into the atmosphere, pollution that has been identified in ice cores.[21]

Unfortunately, we have failed to learn the lessons from ancient Rome. Today, water continues to be contaminated by lead (and other pollutants), such as in Flint, Michigan, and many other places. The sorry story of Flint's lead crisis is told in Gleick's *The Three Ages of Water* and other sources. Briefly, the community was faced with a cutoff of water supply due to excessive lead in the water. Politics—mostly community-state antagonism—has prevented fixing the issue.

Lead poisoning is common among children, seriously harming their learning ability. Much of our freshwater is polluted by agricultural and industrial chemicals, drugs, plastics, sewage, and everything else foul. Alligators and fish are developing major reproductive problems from the amounts of hormones and hormone-like plastic chemicals in the water.

Contamination of water from industry is rapidly increasing everywhere, especially in poorer nations. It includes some horrific problems unknown till recently, including an explosive increase of drugs in the water. Everything from cocaine to birth control pills is contaminating water supplies, with rapidly mounting serious effects.

Another issue is the pollution of groundwater (and so drinking water) from fracking. Fracking is the process by which underground rock is fractured to release trapped natural gas. Unfortunately, the gas becomes mixed with water, which is then pumped out and distributed to business and homes. Stories of flaming water in kitchen sinks have been highlighted in the news.

And Then, Drought

Climate change increases overall rainfall—rain has increased about one percent already and will increase five percent more by 2100. The bad news is that this rain will not be everywhere but will fall largely over the ocean or in already-rainy areas. Thus, despite an overall increase in rain, the dry parts of the world are rapidly getting drier. This is already happening in western North America, Australia, northern Africa, Europe, and the Middle East. Drought conditions reduce water for people, animals, farms, wildlife, and aquatic ecozones, and this results in a reduction of food, loss of livestock, wildfires, famine, and possibly even warfare.

The western US has been in a severe megadrought since 2011, the worst in 1,200 years. (Tree rings tell the story.) In Los Angeles, the lowest recorded rainfall up until 2000 was a year in the 19th century that rained about five inches. Since 2000, Los Angeles received four inches in 2001, three inches in 2006, and two inches in 2007. The winter of 2014–2015 was virtually rainless throughout the whole states of California and Nevada (figures from ongoing daily totals in the *Los Angeles Times*). Projections of enormous rains the following winter were not fulfilled; southern California was drier than ever and reservoirs were approaching empty. The low moisture content of the vegetation set conditions for massive wildfires. Then in 2022–2023, California (and the western US) was hit by "atmospheric rivers" that brought massive amounts of rain and snow, causing major flooding and associated consequences. This has helped to increase the water levels in reservoirs, but not enough to declare the drought being over.

Severe droughts can be catastrophic. In ancient Egypt the failure of the Pharaoh to bring the yearly Nile flood caused revolutions. Some of the Ancient Maya groups succumbed to drought about 1,000 years ago, and droughts have depopulated whole areas of the US, as in the dust bowl in Oklahoma and the Oregon desert. Droughts are also forcing people to migrate to wetter areas, and that is also a problem.

How Did This Happen?

A major issue in all these problems is climate change. But another big one is government mismanagement due to incompetence, corruption, and bureaucratic paralysis. Mismanagement of water resources not only leads to loss of water; it leads to poisoned soil. Salts of all kinds leach down from upstream or leach upward from deep in the Earth. In Australia, ENA observed that considerable farmland has been rendered unusable because farmers cleared off the forest and planted wheat. Without the deep roots of the trees, the groundwater from deep underground moved upward, carrying salt. The ground over many acres of Australia is now whitening with salt and will be unusable for millennia.

Public relations campaigns endlessly "spin" the economic benefits of pollution, the need for rampant and unregulated economic growth, and the inexhaustibility of freshwater and other resources. The western US has been repeatedly fooled by inflated figures, using, for instance, far above-average river flows as baselines.[22] Agriculture has changed from careful management of water to considerable waste, partly due to the rise of big agribusiness.

In Africa, over 54 million acres of land have recently been taken over by foreign governments and firms in a new rush to colonize and control the resources of that hapless continent. These will be farmed, supposedly, and all too often the methods are those of high-tech agribusiness, using enormous quantities of chemicals and water. This will stress the regions in question and deprive the local residents of water.

The poster child for water mismanagement is the Aral Sea, a vast lake in a closed basin in Kazakhstan and Uzbekistan in central Asia.[23] For millennia, the lake was sustained by model water management. Some of this management was developed by unlikely heroes, including Genghis Khan and Tamerlane the Conqueror. A rich agricultural economy producing wheat, barley, silk, melons, vegetables, and livestock developed along the Amu and Syr Rivers and was sustained for thousands of years. There was also a huge, productive, and sustainable fishery in the Aral Sea itself.

The Soviets changed all that, turning the entire basin into a vast cotton field. However, cotton requires far more water than the traditional economy did, and now Uzbekistan and Kazakhstan are among the highest per capita users of water in the world. In addition, cotton uses more chemicals than any other crop; one-third of all the pesticides in the world are used on cotton.

The increased demand for water led to disaster for both the natural and cultural environments. The lake mostly dried up and now is just a salty puddle, some ten percent its original size, as the rivers that once fed it no longer reach it. A toxic mix of natural salts and accumulated pesticides and fertilizers blows over the desert plains. The southern portion of the basin is in Uzbekistan, which is trapped in a vicious circle: it cannot stop growing cotton, the source of most of its income, and cannot make enough from cotton to do much. A good deal of the irrigation water drains off into small temporary lakes that are polluted and erratic.

The once thriving, diverse, and ancient agricultural communities were irreversibly altered with its people forced to become cotton farmers. The knowledge about the traditional management of the lake was lost. The once productive fishery is gone, along with the natural habitats that supported a myriad of other life. A true tragedy.

The north end of the basin is fortunate to be in Kazakhstan, a country that is richer and has a more diverse economy. Kazakhstan has dyked off its end of the basin, which includes the Syr River delta, and is slowly restoring that end of the sea, but there is little hope of a restoration of the rest. The Aral basin is likely ruined forever. It will never produce more than a tiny fraction of the wealth it produced in Tamerlane's time. Meanwhile, as the irrigated lands become ever more salty and poisoned by pesticides, they will soon go out of production permanently. The other rivers that provide water to Turkmenistan and Afghanistan are in a similar situation. How long water supplies will last in these formerly rich lands is an open question.

Also typical is waste of water by the rich for trivial purposes, to the ultimate suffering of everyone. Vast lawns and golf courses take up a huge percentage of the water in urban and developed areas. In California, water withdrawn by giant agribusiness for very low-value agriculture (irrigating wild hay, potatoes, and the like) has destroyed extremely productive and high-value fisheries as well as wetlands that had less quantifiable but no less real values. California now faces a huge water crisis that will send water prices sky-high for everyone—even those who can afford to play golf.

Legal Issues in Water Management

Much of the US operates under some form or other of English common law, variously adapted. This guarantees riparian rights: people on a watercourse have rights to the water. In this context, it is well to remember that the English word "rival" derives from Latin *rivus*, "riverbank," as does the word "river." Twain's famous comment on water in the west, quoted at the beginning of this chapter, emphasizes the point. These "rights" have been widely extended to water allocation, including a "first in time, first in right" principle that is the greatest bane of California water law. Agricultural interests that descend, legally, from those established in the 19th century dominate the state, and they sell, rent, or otherwise profit from their rights.

This leaves groundwater in a legal limbo. Normally, anyone who owns the surface of the land owns the right to pump the water—in striking contrast to the rules concerning oil and minerals. As a result, groundwater aquifers have become common-pool resources and are rapidly being exhausted. Climate change is causing drought in arid parts of the world, including California and the southwestern US, and this has led to massive withdrawals of groundwater—many times more than could be recharged even in a good year, let alone in the horrible droughts that have followed on global climate change. Until 2014, California (unlike other western

US states) had no regulations on groundwater use. In late 2014, the California Legislature finally recognized that catastrophic drought was forcing their hand and was here to stay, so even the most obstructionist Republicans agreed to pass a groundwater regulation law.[24]

Access to water should be one of the most basic matters of environmental justice.[25] However, action on this issue is uncommon, and people cope with the results of injustice as best they can. A team notable for the incredible level of expertise—ranging from water expert Amber Wutich to veteran Puerto Rican anthropologist Carlos García-Quintano—studied the survival of people left without water by Hurricane María in 2017.[26] The US government under Donald Trump did essentially nothing to help, although Trump did pass out some paper towels. This came after decades of neglect by the federal government of Puerto Rico's infrastructure, including its water systems. The team observed cases totaling 20,000 bottles of water sitting neglected in a government facility while Puerto Ricans died for a lack of safe water. Yet people managed through sharing and mutual aid—refuting Hobbes' cynical view that humans lack moral compass without regulation.

Will countries fight over water? In a 2009 article, Wendy Barnaby noted that no modern country has come even close to fighting over water. Unfortunately, since that article, water conflicts have suddenly increased, becoming frighteningly common. There was a sharp rise in conflict in 2012 and an extremely sharp rise in 2018, following droughts in many areas, and water was catapulted into the rank of major world conflicts.[27]

Some countries are truly desperate: Tunisia, Afghanistan, and Jordan, among others. Some are very close to the edge and will not be able to carry out current development plans without extremely major changes in water management; this includes China, India, and Iran. Some are in desperate straits because they are downstream: Egypt, Syria, and Iraq depend almost entirely on river flow from other countries.

Barnaby points out that Egypt has treaties with its upstream suppliers, but those are Sudan and Ethiopia, countries with no history of honoring such scraps of paper. Syria and Iraq are in worse shape since their supplier, Turkey, has a military that could beat both (and several other countries) at once with ease, especially given their current chaotic state. Big dams under construction in Turkey could cut off water to those nations.

Moral Issus in Water Management

It would be very hard to imagine a moral or religious code that denied water to those dying of thirst. Yet, modern governments do exactly that, by wasteful and corrupt development schemes, privatization of water, permitting contamination, displacement of impoverished people, and many other practices.

Gary Chamberlain, in a valuable book titled *Troubled Waters*,[28] reviewed the status of water in all the world's major religions and found that all of them are quite specific about enjoining us to treat water as a common good to share with all who need it. Certainly, of all human needs, water is second in immediate importance only to oxygen. Water is needed every day, in fairly large quantities, by every human. It is needed directly for drinking and washing, indirectly in much greater quantities for food production and manufacturing. It is irreplaceable; the economists' notion of "infinite substitutability" breaks down totally here. Water must be reasonably pure to be useful—the purer the better.

Some of the most interesting research on water in the Middle East refers to Muslim or Muslim-influenced local irrigation systems. This is in large part because Muslim law, developed in arid lands, is quite specific about water. Gary Chamberlain, synthesizing a number of sources, reports: "Muslim law codes... forbid private ownership of water, at least in its natural state. There is a hierarchy of uses...first is the right of thirst...no one can be denied the water necessary to drink...then all are allowed water for their daily needs of bathing, cleaning, cooking, and so forth." Then "next comes the right to provide water to livestock; and last comes the irrigation of crops, which consumes the most water. Only when water has been placed in a vessel...is water considered a private good."[29]

This is a priority partly because Islam enjoins cleanliness, making thorough washup and bathing a religious duty. However, even ritual cleaning must not be wasteful. Muhammad, according to one account, once saw his early follower Sa'ad "performing the ablutions...using a lot of water, he intervened, saying: 'What is this? You are wasting water.' Sa'ad replied asking: 'Can there be wastefulness while performing the ablutions?' To which God's Messenger replied: 'Yes, even if you perform them on the bank of a rushing river.'"[30]

"Water distribution has very clear-cut legislation in Islam. In general terms its rules are based on the principle of benefiting all those who share its watercourse."[31] The duty to provide water for livestock is taken very seriously, Islam having originated among desert travelers. Accounts describe careful management of flocks at the wells, with the most water-needing animals drinking first.

This emphasis on common property led to intricate but efficient and enforceable common property regimes being established in Muslim lands. One significant case is in the Sierra de la Contraviesa area southeast of Grenada, studied by Gaston Remmers, among others.[32] Remmers describes an incredibly sophisticated system for making sure that everybody has fair access to irrigation water, no matter how wet or dry the year. The village social organization is based on water management. (Spain has other successful irrigation systems without obvious Moorish ancestry, too; see endnote 50 in Chapter 2.)

Islamic management adopted in Spain was then transmitted to the New World, where it survives in remote Hispanic and Native American communities from New Mexico and Sonora to Peru and Argentina. The Quechua of Peru, already famous for their spectacular canals often cut out of solid rock, developed irrigation societies complete with annual fiestas, water allocations, and such strict morals that people not only protected each other's water, they successfully resisted the Peruvian government when it tried to take water from villages in the Colca Valley.[33]

Another system maintained by a religious organization is in Bali. Stephen Lansing studied this system over many years.[34] Irrigation on that Indonesian island is derived from water coming from the lake in the crater of a single gigantic inactive volcano at the top of the island. The water is sacred. The head priest of the island, the *jero gde*, lives at the lake outlet, and oversees the water system and is appointed more for his hydrological expertise than for his religious devotions. A hierarchy of priests, progressively farther and farther downstream, oversees the breakup of this stream into tens, hundreds, and finally tens of thousands of channels. These feed a vast system of rice paddies; the island is one huge farm, growing mainly rice but also dozens of tropical crops. Water is timed so that there is no single pulse of irrigation. That would not only take too much water; it would allow insect pests to multiply out of control. Instead, each field has its own schedule of irrigating and drying off. The World Bank came in with sophisticated technology in the late 1980s to improve this system and promptly caused disaster. Their computer-assisted plans led to water shortages, local floods, and insect outbreaks. Control promptly went back to the *jero gde*. Lansing modeled the traditional system with his own computers and found it to be about as perfect as could be achieved in the real world.

Similar, if less comprehensive and perfect, local systems of terracing and water control are well documented from elsewhere in Indonesia, as well as from the Philippines, pre-American Hawaii, New Guinea, and indeed most of Oceania and the rest of the montane tropical and subtropical world. Usually, religion is marshaled to help maintain them. Often, they are also maintained through kinship systems, as in Luzon and among the Toba Batak of Sumatera (studied by ENA's former student, the late Richard Lando, in the 1970s). Often, they produce fish and other animal proteins as well as staple plant foods. India has countless religiously maintained irrigation systems too (a particularly superb account is by David Mosse[35]).

Common property management works in today's world. Native Americans are reviving some of their traditional management techniques in the arid Southwest. Modern societies are not necessarily bad managers, and some succeed at least as well as traditional people.

Elinor Ostrom's original studies of successful common property management were done in the Los Angeles Basin.[36] She found that the dozens of cities sharing the basin had been forced to work together to manage the small rivers that provide water and carry away sewage. She compared this management with that of the Mojave River, just outside the Los Angeles Basin. Here, powerful mining interests control the headwaters. Next downriver are the relatively well-off towns of Hesperia and Victorville. The river dies, except during the occasional flood, in the desert just past Barstow. This unfortunate town, always poor, has become poorer and poorer as its water source is sucked away. Having less and less political-economic clout, it progressively loses to the mines and the richer towns. Barstow is slowly strangling to death.

It is, indeed, hard to avoid worshiping water. If one has any religious regard for nature. One of the striking facts about humans is that, everywhere, they seem to honor and revere waterfalls. Major falls are parks and pilgrimage spots in the US, China, and elsewhere. Traditional small-scale societies everywhere seem to have worshiped them. The sheer force of the water at such points is hypnotizing. One can stand looking in a sort of trance for minutes or hours. Lakes and deep pools, and above all the vast ocean, have a different kind of spiritual sense: peaceful and calm, yet evidently extremely powerful. The power is latent. One knows that a storm or a break in a water barrier could unleash it at any moment.

Even if one is not religious, any concern for anything outside one's own narrowest self-interest simply must include concern for water. The future for all of us is bleak unless immediate action is taken on a global scale. A major work on the ethics of water, *Ethical Water Stewardship*, edited by Ingrid Stefanovic and Zafar Adeel,[37] covers a whole range of viewpoints: religious, Indigenous, and philosophical. They emphasize the "Platinum Rule": Do unto others what *they* want or need, not what you want (as in the Golden Rule). They confirm that religions, especially Near Eastern ones, make a point of mandating water to the thirsty, and Indigenous traditions everywhere direct us to revere and care for this precious substance.

What to Do?

We cannot easily get more water. Towing Antarctic icebergs north and desalting sea water are the only large-scale possibilities, and they are both expensive. In reality, far more efficient use of water is the only solution, and it is imperative. Few people seem to realize that human use of water for drinking and bathing is utterly insignificant relative to the use of water in agriculture and industry. Personal use is less than one percent of all water use.

The rich use much more water than the poor, since rich people like huge lawns and water-sucking landscaping, as well as large swimming pools and golf courses. In California overall, households and their outdoor landscaping use an average of 360–400 gallons per day. Poor neighborhoods tend to have apartments in areas lacking much landscaping but covered instead with asphalt and concrete. Thus, the rain that does fall flows directly into the ocean instead of replenishing groundwater. Thus, ironically, more water is lost than they consume.

Americans love their lawns and landscaping, but these are water sinks. The use of dry scaping using drought-tolerant plants and natural landscaping in place of grass lawns is becoming more popular. The total area of lawns and related water-demanding landscaping in the US is greater than the area of Pennsylvania. They take a wholly disproportionate amount of water, pesticides, and fertilizers. Other targets of water conservation are more efficient sewage treatment and the control of golf course landscaping. Every drop counts.

Most of the water is used by agriculture, and inefficient irrigation methods, poor crop choices (recall the cotton in the Aral Sea), and demand for high-water usage products (e.g., beef and milk) result in a huge waste of water. The use of drip irrigation instead of sprinklers or flooding would help. For example, the production of a single almond requires 1.1 gallons of water, a head of broccoli takes 5.4 gallons, an orange 13, a single walnut 4.9, a tomato 3.3, a strawberry 1.9, and many of these products are 100 percent irrigated. It gets worse; it takes 230 gallons to produce one pound of rice, 200 gallons for a pound of beans, 130 gallons for a pound of wheat, 50 gallons for a pound of corn, 410 gallons for a pound of chicken, 1,000 gallons for a pound of pork meat, and 1,500 gallons for a pound of beef, with cotton being in the range of pork and beef.[38] This is a huge waste of water.

We could eat less (most Americans eat too much anyway) and waste less, and/ or we could start eating foods that required less water to produce. There are two particularly fine books about how to do dryland farming: David Cleveland and Daniela Soleri's *Food from Dryland Gardens* and Gary Nabhan's *Growing Food in a Hotter, Drier Land.*[39]

The production of meat and milk is particularly water consumptive. Most cows are raised on feed grown on irrigated fields and demand huge amounts of water for drinking and washing; then processing their meat and milk takes yet more water. A reduction in the consumption of meat and milk would help in several ways: less water, fewer animals killed, and maybe some weight loss. If we do not want to eat good foods such as cactus fruit, mesquite beans, and prickly pear pads, we could at least feed them to the animals and save the grains for people.

We could save water in other ways. One way is to take into account the water needed to produce the various agricultural commodities and manufactured goods. For example, if a person bought a cotton shirt, they are probably not using any American water to speak of, using instead enormous quantities of Egyptian and Chinese water—assuming, as if often the case, that the cotton is grown in Egypt and the shirt is made in China. The horrific case of growing cotton in Uzbekistan is entirely export driven. Whoever gets the good-quality cotton items made from Uzbeki cotton is ruining that desperately stressed nation but is probably quite unaware of it.

On the other hand, Wendy Barnaby[40] pointed out that a dry country can spare its limited water resources by importing food and not growing crops with high water requirements. Most of the water used to produce grain and coffee comes from rainfall, but most of the water used to produce meat, cotton, and many vegetables is from irrigation or piped water. Thus, by eating lower on the food chain and wearing clothing more economically, we could save enormous amounts of water.

Another unconventional solution comes from Peru. Located on the Pacific coast, Lima is essentially rainless but very foggy (especially in winter) because of the cold water of the Humboldt Current just offshore. In ancient times, this fog sustained lomas—areas of dense vegetation, even forests, inhabited by animals as large as deer, but these forests were later largely destroyed by overgrazing, wood cutting, and mining. Today, there is an attempt to restore the lomas. Large, dense nets have been set up, on which the fog forms water droplets that are directed down to young trees. When the trees are older, they will strain their own fog, thus restoring the old lomas forests.[41] This idea could be used in many other places where cold currents run along desert shores: Baja California, Morocco, South Africa, and elsewhere. Similar in concept, another idea is to build large vapor towers in the ocean to "harvest" atmospheric freshwater.

Desalinization is an alternative source of water and has been used in many coastal countries both for domestic and agricultural purposes. Israel, Saudi Arabia, Australia, and others are dependent on this water source and its production and use are rapidly growing. However, at this time, the process is expensive, but better technology is rapidly developing and may yet save us.

Notes

1 Yamada, Yosuke, et al. 2022. "Variation in Human Water Turnover Associated with Environmental and Lifestyle Factors." *Science* 378:909–915.
2 Pearce, Fred. 2007. *When the Rivers Run Dry: Water—the Defining Crisis of the 21st Century*. Boston, MA: Beacon; Gleick, Peter. 2023. *The Three Ages of Water: Prehistoric Past, Imperiled Present, and a Hope for the Future*. New York: Public Affairs.
3 Harrell, Stevan. 2023. *An Ecological History of Modern China*. Seattle: University of Washington Press. Zheng, Chunmiao, and Zhilin Guo. 2022. "Plans to Protect China's Depleted Groundwater." [Letter.] *Science* 375:827.
4 Drake, Nick A., Roger M. Blench, Simon J. Armitage, Charlie S. Bristow, and Kevin H. White 2011. "Ancient Watercourses and Biogeography of the Sahara Explain the Peopling of the Desert." *Proceedings of the National Academy of Sciences* 108(2):458–462.
5 Bengali, Shashank, and Ramin Mostaghim. 2018. "In Iran, an Environment for Unrest." *Los Angeles Times*, Jan. 18, A3; Bulos, Nabih. 2022. "In Once-Fertile Iraq, a Way of Life Dries Up." *Los Angeles Times*, Jan. 1, A1, A4; Bulos, Nabih, and Omid Khazani. 2021. "Tehran, and Much of Iran, Slowly Sinking." *Los Angeles Times*, Aug. 2, A3; Bulos, Nabih, and Marcus Yam. 2021. "An Ancient Valley Facing Extinction." *Los Angeles Times*, A1, A4–A5; Jaleh, Babak, and Mahtab Eslamipanah. 2023. "Restore Iran's Declining Groundwater." [Letter.] *Science* 379:148.
6 Wittfogel, Karl. 1957. *Oriental Despotism*. New Haven, CT: Yale University Press; also see Price, David H. 1994. "Wittfogel's Neglected Hydraulic/Hydroagricultural Distinction." *Journal of Anthropological Research* 50(2):187–204.
7 Katz, David. 2006. "Going with the Flow: Preserving and Restoring Instream Water Allocations." In *The World's Water 2006–2007: The Biennial Report on Freshwater Resources*, Peter Gleick et al. (eds.), pp. 29–49. Washington, DC: Island Press; Pekel, Jean-François, Andrew Cottam, Noel Gorelick, and Alan S. Belward. 2016. "High-resolution Mapping of Global Surface Water and Its Long-term Changes." *Nature* 540:418–422.
8 Harrell, Stevan. 2023. *An Ecological History of Modern China*. Seattle: University of Washington Press; Stone, Richard. 2008. "Three Gorges Dam: Into the Unknown." *Science* 321:628–632.
9 Peterson, Maya. 2019. *Pipe Dreams: Water and Empire in Central Asia's Aral Sea Basin*. Cambridge: Cambridge University Press. Richter, Brian D., et al. 2024. "New Water Accounting Reveals Why the Colorado River No Longer Reaches the Sea." *Communications Earth & Environment* 5, article 134.
10 Harrell, Stevan. 2023. *An Ecological History of Modern China*. Seattle: University of Washington Press.
11 For this and following paragraph, see Vörösmarty, Charles J., Peter B. McIntyre, Mark O. Gessner, David Dudgeon, Alexander Prusevich, Pamela Green, Stanley Glidden, S. E.

Bunn, C. A. Sullivan, C. Reidy Liermann, and P. M. Davies. 2010. "Global Threats to Human Water Security and River Biodiversity." *Nature* 467:555–561.

12 Hovey, Tim E. 2001. "When Nature Finds a Way: Return of the Southern Steelhead." *Outdoor California*, March-April, 17–20.

13 Glennon, Robert Jerome. 2004. *Water Follies: Groundwater Pumping and the Fate of America's Fresh Waters*. Washington, DC: Island Press; Kuhn, Eric, and John Fleck. 2019. *Science Be Dammed: How Ingoing Inconvenient Science Drained the Colorado River*. Tucson: University of Arizona Press; Wheeler, Kevin G., et al. 2022. "What Will It Take to Stabilize the Colorado River?" *Science* 377:373–375.

14 Abbott, David R. (ed.). 2003. *Centuries of Decline during the Hohokam Classic Period at Pueblo Grande*. Tucson: University of Arizona Press; Rea, Amadeo. 1983. *Once a River: Bird Life and Habitat Changes on the Middle Gila*. Tucson: University of Arizona Press; Redman, Charles L. 1999. *Human Impact on Ancient Environments*. Tucson: University of Arizona Press; Russell, Frank. 1975. *The Pima Indians*. Tucson: University of Arizona Press.

15 James, Ian. 2023. "The Shrinking Colorado River." *Los Angeles Times*, July 31, A1, A6.

16 https://www.ladwpnews.com/ladwp-achieves-99-4-dust-control-reduction-at-owens-lake/.

17 Laylander, Don. 1997. The Last Days of Lake Cahuilla: The Elmore Site. *Pacific Coast Archaeological Society Quarterly* 33(1–2):1–138.

18 Frie, Alexander L., Alexis C. Garrison, Michael V. Schaefer, Steve M. Bates, Jon Botthoff, Mia Maltz, Samantha C. Ying, et al. 2019. "Dust Sources in the Salton Sea Basin: A Clear Case of an Anthropogenically Impacted Dust Budget." *Environmental Science & Technology* 53(16):9378–9388.

19 Daigle, Katy. 2016. "Death in the Water." *Scientific American*, Jan., 42–51.

20 A discussion of this issue is provided in Cilliers, Louise, and Francois Retief. 2019. "Lead Poisoning and the Downfall of Rome: Reality or Myth?" In *Toxicology in Antiquity* (2nd ed.), Phillip Wexler (ed.), pp. 221–229. London: Academic Press.

21 McConnell, Joseph R., Andrew I. Wilson, Andreas Stohl, Monica M. Arienzo, Nathan J. Chellman, Sabine Eckhardt, Elisabeth M. Thompson, A. Mark Pollard, and Jørgen Peder Steffensen. 2018. "Lead Pollution Recorded in Greenland Ice Indicates European Emissions Tracked Plagues, Wars, and Imperial Expansion during Antiquity." *Proceedings of the National Academy of Sciences* 115(22):5726–5731.

22 Glennon, Robert Jerome. 2004. *Water Follies: Groundwater Pumping and the Fate of America's Fresh Waters*. Washington, DC: Island Press; Kuhn, Eric, and John Fleck. 2019. *Science Be Dammed: How Ingoing Inconvenient Science Drained the Colorado River*. Tucson: University of Arizona Press.

23 The Aral Sea story is well covered. See Kobori, Iwao, and Michael H. Glantz (eds.). 1998. *Central Eurasian Water Crisis: Caspian, Aral, and Dead Seas*. Tokyo: United Nations University Press; Micklin, Philip, and Nikolay V. Aladin. 2008. "Reclaiming the Aral Sea." *Scientific American*, April, 64–71; Varis, Olli. 2014. "Curb Vast Water Use in Central Asia." *Nature* 514:27–29.

24 On this general issue see Gies, Erika. 2022. *Water Always Wins: Thriving in an Age of Drought and Deluge*. Chicago: University of Chicago Press.

25 Palaniappan, Meena, Emily Lee, and Andrea Samulon. 2006. "Environmental Justice and Water." In *The World's Water 2006–2007: The Biennial Report on Freshwater Resources*, Peter Gleick et al. (eds.), pp. 117–144. Washington, DC: Island Press; Stefanovic, Ingrid Leman, and Zafar Adeel (eds.). 2020. *Ethical Water Stewardship*. Cham, Switzerland: Springer; Wutich, Amber, and Alexandra Brewis. 2014. "Food, Water, and Scarcity: Toward a Broader Anthropology of Resource Insecurity." *Current Anthropology* 55:444–468.

26 Roque, Anais, Amber Wutich, Alexandra Brewis, Melissa Beresford, Hilda Lloréns, Carlos García-Quintano, and Wendy Jepson. 2023. "Water Sharing as Disaster Response: Coping with Water Insecurity after Hurricane María." *Human Organization* 82:248–260.

27 Barnaby, Wendy. 2009. "Do Nations Go to War over Water?" *Nature* 458:282–283; Gleick, Peter. 2023. *The Three Ages of Water: Prehistoric Past, Imperiled Present, and a Hope for the Future*. New York: Public Affairs.

28 Chamberlain, Gary L. 2008. *Troubled Waters: Religion, Ethics, and the Global Water Crisis*. Lanham, MD: Rowman and Littlefield.

29 Chamberlain, Gary L. 2008. *Troubled Waters: Religion, Ethics, and the Global Water Crisis*. Lanham, MD: Rowman and Littlefield, p. 54.

30 Özdemir, Ibrahim. 2003. "Toward an Understanding of Environmental Ethics from a Qur'anic Perspective." In *Islam and Ecology: A Bestowed Trust*, Richard C. Foltz, Frederick M. Denny, and Azizan Baharuddin (eds.), pp 3–38. Cambridge, MA: Harvard University Press for the Center for the Study of World Religions, Harvard Divinity School.

31 Dien, Mawil Izzi. 2003. "Islam and the Environment." In *Islam and Ecology: A Bestowed Trust*, Richard C. Foltz, Frederick M. Denny, and Azizan Baharuddin (eds.), pp. 107–119. Cambridge, MA: Harvard University Press for the Center for the Study of World Religions, Harvard Divinity School.

32 Remmers, Gaston G. A. 1998. *Con Cojones y Maestría*. Amsterdam: Thela; for other successful systems in Spain, see Grove, A. T., and Oliver Rackham. 2001. *The Nature of the Mediterranean World*. New Haven, CT: Yale University Press; Guillet, David. 2006. "Rethinking Irrigation Efficiency: Chain Irrigation in Northwestern Spain." *Human Ecology* 34:305–329.

33 Gelles, Paul H. 1995. "Equilibrium and Extraction: Dual Organization in the Andes." *American Ethnologist* 22:710–742; Gelles, Paul H. 2000. *Water and Power in Highland Peru: The Cultural Politics of Irrigation and Development*. New Brunswick, NJ: Rutgers University Press. See also Trawick, Paul. 2001a. "The Moral Economy of Water: Equity and Antiquity in the Anean Commons." *American Anthropologist* 103:361–379; Trawick, Paul. 2001b. "Successfully Governing the Commons: Principles of Social Organization in an Andean Irrigation System." *Human Ecology* 29:1–25; Trawick, Paul. 2002. *The Struggle for Water in Peru: Comedy and Tragedy in the Andean Commons*. Stanford: Stanford University Press.

34 Lansing, Stephen. 1987. "Balinese 'Water Temples' and the Management of Irrigation." *American Anthropologist* 89:326–341; Lansing, Stephen. 1991. *Priests and Programmers: Technologies of Power in the Engineered Landscape of Bali*. Princeton, NJ: Princeton University Press; Lansing, Stephen. 2006. *Perfect Order: Recognizing Complexity in Bali*. Princeton, NJ: Princeton University Press.

35 Scarborough, Vernon (ed.). 2003. *The Flow of Power: Ancient Water Systems and Landscapes*. Santa Fe: SAR Press; Mosse, David. 2006. "Rule and Representation: Transformations in the Governance of the Water Commons in British South India." *Journal of Asian Studies* 65:61–90; Mosse, David. 2003. *The Rule of Water: Statecraft, Ecology, and Collective Action in South India*. New Delhi: Oxford University Press.

36 Ostrom, Elinor. 1990. *Governing the Commons: The Evolution of Institutions for Collective Action*. New York: Cambridge University Press.

37 Stefanovich, Ingrid Leman, and Zafar Adeel. 2020. *Ethical Water Stewardship*. Cham, Switzerland: Springer.

38 Park, Alex, and Julia Lurie. 2014. "It Takes How Much Water to Grow an Almond?" *Mother Jones*, Feb. 24, online, http://www.motherjones.com/environment/2014/02/wheres-californias-water-going.

39 Cleveland, David, and Daniela Soleri. 1991. *Food from Dryland Gardens: An Ecological, Nutritional, and Social Approach to Small-Scale Dryland Food Production*. Tucson: Center for People, Food and Environment; Nabhan, Gary Paul. 2013. *Growing Food in a Hotter, Drier Land: Lessons from Desert Farers on Adapting to Climate Uncertainty*. White River Junction, VT: Chelsea Green.

40 Barnaby, Wendy. 2009. "Do Nations Go to War over Water?" *Nature* 458:282–283.

41 Vince, Gaia. 2010. "Out of the Mist." *Science* 330:750–751.

7

THE FARMS AND FIELDS THAT FEED US

"The nation that destroys its soil, destroys itself." (Franklin D. Roosevelt)

For millions of years, all people got their food and most everything else from hunting and gathering wild plants and animals. Such societies were generally simple socially and politically, with small populations and relatively simple technology. While these hunter-gatherers impacted their environment, the scale of that impact was very small. Cutting down a tree or clearing an acre or two of forest is a very minor trauma, one that was easily restored with no lasting damage.

Then, about 12,000 years ago, people began to domesticate a few of the wild plants and animals they had been exploiting by slowly gaining control of their genetics and manipulating their reproduction to produce forms better suited to humans. The first domestications took place in the Near East. China and the Western Hemisphere independently domesticated grains not long after. The development of productive and varied agriculture was a stunning achievement, taking millennia to complete, but revolutionary in final effect. Agaves, for instance, were domesticated late in pre-Columbian Mexico and the American Southwest, diversifying agriculture there still more.

Grains were made more productive and easier to harvest. Sheep were made dumber so they could be herded—domestic sheep have only about three-quarters as much brain as wild ones and were bred for wool, an unnatural product developed only after millennia of domestication. Cattle were bred to increase meat and milk production. Horses were bred for riding and for pulling carts.

Still, people continued to hunt and to gather wild species. Most contemporary farmers still do some hunting and gathering; in fact, it remains a critical aspect of many

DOI: 10.4324/9781003560326-7

economies. Even in industrialized societies, the use of wild resources continues, albeit a very small aspect of the economy. Anyone who has been fishing, hunting, or gathering wild berries is a "hunter-gatherer."

Farming eventually supplanted hunting and gathering in almost all societies, with a few small exceptions. Agriculture has been with us for over 12,000 years and is a human-made niche, or series of niches.[1] Animal domestication was part of it from the beginning. For a long time, farming, or food production, improved life (although there is disagreement on that point), but the system has reached a crisis point.[2]

A vast range of crops was developed through domestication involving selecting plants and animals for desirable traits. Densely populated areas all learned to have a major starch staple (wheat and barley in the Near East, corn in Mexico, potatoes in the Andes, and so on). They also found vegetable sources of protein, usually species of the bean family. They also grew a range of crops rich in vitamins and minerals; although unaware of the specific chemistry, people still realized that their children did not thrive without including such foods to the starches.

Many of the best crops for future food and sustainable farming are now neglected because they are associated with poorer and less affluent parts of the world. Africa has been particularly shorted, despite a whole range of crops especially well adapted to its hot climate and often poor soils.[3]

Thus, virtually everyone is now dependent on farmed foods. Farming has evolved from very small plots requiring substantial human labor that produced food for the family to larger farms requiring substantial animal labor that produced some surpluses, to the contemporary Western industrial systems of huge farms requiring very substantial mechanical labor along with massive amounts of chemicals and fossil fuels but that produce huge surpluses. In the US today, only about three percent of the population are farmers, and most of these are very small operators—often a retired couple clinging to the family farm. Only one percent of the American population produces large amounts of food; of course, a great deal of our food is imported, but we export even more, especially corn and soybeans. In the rest of the world, the productivity of farming varies from high to low.

There is currently enough food grown to feed the people on the planet, but shortages in specific areas abound. There are food emergencies in a number of places, but these are not due to a shortage of food in general. They are caused by political issues such as the war in Ukraine impeding the shipment of grain and the war in Sudan preventing the delivery of food. Waste of food is also involved; up to a third of world food is lost along the pipeline, from stored harvests to plate waste.

This begs the question: how many people can the Earth support? The current (2023) population is about eight billion people, the bulk of whom live in south and east Asia (India, China, and other countries). Can farms produce more? How much more? When will the system crash? Are we smart enough and resilient enough to adjust?

Contemporary Industrialized Farming

The Western system of industrialized farming is highly complex, extraordinarily productive, quite destructive of biodiversity, dependent on fossil fuel and chemicals, water wasteful, and ultimately unsustainable. Because it is more productive in the short term, it is expanding and replacing sustainable traditional systems that are less productive.

This industrialized agricultural system is relatively new, really coming into its own after World War II. This system is large-scale and highly specialized. Farms are now quite large (to the detriment of small family farms) and focus on a relatively few species, with corn, wheat, and rice being the most important of the crops worldwide. Fields are created across the natural landscape, replacing a much more diverse natural system, with a resulting loss of biodiversity. Often, a single species, such as corn, is grown in the field (monoculture). This lack of genetic diversity makes disease and pests a major issue. Contagion is maximally likely.

In this system, fields are intensively managed to maximize crop yield. To do this, chemical fertilizers are liberally applied to maintain fertility. This comes at the cost of polluting the groundwater and surface runoff that eventually flows to the ocean to create dead zones. But a farmer in Kansas does not think about the dead zone in the Gulf of Mexico caused by overfertilization; it is someone else's problem. Even fertilized fields will lose fertility, requiring even more chemicals. Soil erosion is another issue that is even more difficult to solve, though no-till agriculture and cover crops now greatly reduce it in many areas. Pesticides and herbicides are used to control pests and weeds (a weed is any plant growing where you do not want it, so your prized flower may be someone else's weed).

Farmland is being lost rapidly to desertification, urbanization, erosion, and pollution. In *Harvesting the Biosphere*, Vaclav Smil points out that people settle in the best farming areas, then expand their settlements so that the best soil is the first to be paved over.[4] This process has cost California a third of its best farmland in the last 200 years. China and India, with their extremely fertile but limited good soil and their dense populations, are particularly damaged. China has terribly polluted its land with chemicals. In Canada, the Vancouver urban area is rapidly paving over the only area in western Canada that has good soil and a warm climate. By 2030, the decline of available farmland will be seriously affecting production in Africa and Asia.[5]

Erosion has impacted farming for millennia. It had a share in the decline of Rome and other ancient societies.[6] About a third of the topsoil of the Midwest has washed away since 1700.[7] Erosion is now well controlled there, so well that the Mississippi Delta is starved for new sediment, but erosion is still problematic locally. Elsewhere in the world, the situation is grimmer. China continues to lose much of its soil to wind and water erosion. Farmland will continue to expand for some decades, especially as tropical and temperate forests are converted to agriculture, but this farmland is largely inferior in quality, and it may not be enough to offset loss to desertification and erosion.

A subproblem is sand mining.[8] Vast quantities of sand are necessary to make concrete for construction. Sand is proverbially abundant, but in today's super-populous world, it is depleted by necessary infrastructure building. Rivers and farmland are being sacrificed. One can truly say the sands of time are running out.

Water for this system was, until recently, available in sufficient quantities (at least in the US), so there was little incentive to use it efficiently. Fields are flooded or irrigated with large sprinklers, with too much of the water intended for the plants lost to evaporation. Drip irrigation and other water-sparing methods are rapidly increasing in use, but the bad old ways continue in too many places. Further, irrigation introduces salt and other substances to the soil that, in some cases, will soon make it impossible to grow anything but saltbush. In the Central Valley of California, many thousands of acres of former farmland have reverted to the desert it once was. How much longer can this go on? We do not know.

The vast majority of labor in this system is provided by machines. New machines to replace human workers continue to be introduced. These machines require maintenance and eventual replacement, and a vast and costly infrastructure to meet these needs has developed. Further, the machines run on fossil fuels (gasoline and diesel) that require additional infrastructure and massively contribute to GHG emissions.

However, there are some crops (e.g., radishes, coffee, baby corn, broccoli, and cabbages, to name a few) that must be harvested by hand. Thus, transient farm labor is necessary to feed America, but the system that regulates this labor (often "illegal" farm workers) is broken, thanks to conflicts over the amount of immigration we can accept. Similar problems exist in other countries. This hampers production and raises food prices.

Once harvested, crops are placed in large-scale storage facilities, some refrigerated. This allows the less perishable crops to be stored for years and so mitigate fluctuations in productivity due to drought or other conditions. Yet another complex infrastructure for the transportation and trade of the crops allows them to be moved around the world with little technical difficulty but often with great political difficulty (think of the effects of the Russo/Ukrainian war on world grain trade). All of this, of course, requires still more machines, such as trucks, trains, and ships, and the fossil fuels they need to move.

The dependence on fossil fuels is illustrated by the fact that it takes about 300 calories (a measure of energy) of fossil fuel to produce 100 calories of food, including the energy expended in packaging, transportation, and refrigeration.[9] If the supply of fossil fuels became too costly or difficult to obtain (such as a war in the Middle East), the entire system could collapse (the rise in the price of oil and the economic events of 2008 illustrate this danger). It would be wise to eat foods grown organically and/or locally (visit your nearby farmers' market) to increase efficiency and decrease transport costs. Using biological pest controls would help to lessen the pesticides used by having "good" bugs eat the "bad" bugs rather than killing all the bugs.

On top of growing crops, this system has industrialized the production of animals for their meat, milk, eggs, and leather. Some animals are raised on pasture lands, eating grass that people cannot eat. This is reasonable and efficient. However, most animals are raised in large facilities containing thousands of individuals, sometimes with little room to even move about. Most beef cattle are raised in "feed lots" (Figure 7.1), where they are fed grains and other foods, fattened up, and slaughtered. Dairy cattle are also fed grains and other foods and milked daily. Chickens are raised in vast warehouses with tens of thousands, if not hundreds of thousands, of individuals. These facilities are havens for diseases such as bird flu. One can occasionally read about such outbreaks and the resultant slaughter of millions of chickens to stop the spread of the infection.

The problem with this, apart from animal rights issues, is that about one-third of all agricultural land is used to produce feed for livestock, land that could be much better used to grow food directly for humans. In addition, livestock use vast amounts of water, a resource in short supply. This system also requires a vast infrastructure powered by fossil fuels. Substituting animal feed made from insects such as fly larvae, mealworms, and other larvae, already in production as livestock feed, would improve efficiency.

While contemporary industrial agriculture is highly productive, it is also extremely inefficient, wasteful, and polluting. Chemical fertilizers, herbicides, and pesticides pollute water supplies and unintentionally kill many other plants and animals. The system also results in the large-scale alteration of habitat and in the loss of biodiversity. Further, the system is ultimately unsustainable. This issue is explored in greater detail in the documentary films "Food, Inc." and "Food Inc. 2."

FIGURE 7.1 A feed lot in Saskatchewan, Canada. Photo by Kristen112211 (Wikimedia Commons, Public Domain).

However, the expanding human population essentially requires that, for now, we continue with this system and even expand it. Like drug addicts, we are now "hooked" on this productivity.

Most societies in the past came up with more or less sustainable ways of making a living, supported by ideology.[10] The most efficient food system in terms of supporting more people with a better life while doing less environmental damage is the rice system of East Asia, which extended in historic times to Southeast Asia, where it was perfected. Next come other rice systems, especially India's, which grades into Southeast Asia's.

Some early agricultural states developed different approaches to farming and labor—the difference between large plantations worked by enslaved people versus small privately owned farms. In early Mesopotamia and Egypt, giant plantations worked by servile labor developed early and climaxed in the vast enslaved-worker estates of ancient Rome and Byzantium. In contrast, China moved over time from large plantations worked by enslaved labor to increasing dependence on small farms.

Recent plantations based on slavery in colonial Indonesia, India, Africa, the Caribbean, and elsewhere have gone with monoculture farming, authoritarian regimes, and reactionary politics. The US and Brazil represent cases where slavery and monoculture agriculture (cotton and sugar) have carried forward to affect society ever since.

However, small-scale farming is not a perfect cure: it did not prevent land degradation, nor did it prevent authoritarian rule. China had authoritarian imperial rule throughout its history, today being more extreme than in previous times.

Rights are critical. Traditional societies afford rights of various sorts to the environment; these are not specifically codified but are culturally shared. Today, giving rights to the environment is becoming widespread but remains thin.[11]

Colonialism spread Western-style agriculture around the world. International trade had always involved a flow of edible and wearable goods from the margins to the centers of trade, but this became a worldwide enterprise by 1800, and today the world's less affluent countries depend on exporting raw materials. Even some affluent countries, including the US, are heavily dependent on exporting food. This often goes with legal systems that heavily favor large-scale producers and importers and strongly disadvantage small-scale producers. Giant plantations prosper, often with heavy subsidies, while small farmers suffer. This often involves outright displacement of local people or ecological damage to their homes such that they are forced away.[12]

Today, 38 percent of the land surface of the Earth is devoted to producing food for people.[13] However, most of that acreage is dedicated to grazing, a generally low-intensity and low-productivity endeavor. Production of other goods also impinges on food production. Oil and gas production has taken six percent of wheat land, a similar percentage of rangeland, and less but significant amounts of other productive land in recent years.[14] Mining takes land and pollutes rivers. Various types of agriculture compete with each other, and usable farmland has essentially "peaked." If world farmland begins to shrink in total area, economic and social conflict will rapidly increase.

A relatively new approach is to farm highly salt-tollerant plants for food or fodder. This "saltwater" farming could utilize lands unsuitable to conventional agriculture, although it would damage the natural saltwater environments.

The World Food Problem

As of 2022, a global report issued by the UN with the EU and humanitarian organizations found that 258 million people had dangerous food insecurity. Particularly problematical were the situations in Afghanistan, Burkina Faso, Haiti, Nigeria, Somalia, South Sudan, and Yemen,[15] places where there were ongoing civil wars or local violence. They display a characteristic pattern: despotic rule unable to maintain control. The result ranged from total collapse of functional government in several to malfunctioning government in others. Nigeria is a special case, with violence and want largely confined to the unsettled north.

These countries are not exceptions; they are the future unless immediate measures are taken. The steady decline in resources per capita is one of several factors leading to a rise in authoritarian and tyrannical governments worldwide. Such governments invariably back traditional and rural interests. They fight progressive change, including the changes necessary for coping with the resource crunch and the climate crisis. Also, the dictators usually come to surround themselves with flatterers while being more and more prone to imprison or execute critics.[16] Thus bad policies are established, and the country falls into chaos.

Hong and colleagues[17] show that export agriculture continues to drive climate change and the loss of biodiversity. Interesting is the very low productivity of agriculture in some countries, with Madagascar producing the fewest calories per unit of land. This is due to the extensive cattle-raising and technologically very simple shifting agriculture used across most of that island.

Diversity in fields, on farms, and in the national agricultural scene increases production and stability. Monoculture agriculture is the opposite, less productive and riskier. One recalls the great potato famine of 1846–1848 in Ireland and continental Europe, caused by overreliance on one variety of one plant, as described in Redcliffe Salaman's classic book *The History and Social Influence of the Potato*.[18]

Fertility is another concern. Soil fertility is rapidly exhausted by crops, especially heavy feeders like corn. Nutrients in soil are highly mobile, rather than stable over time, and can go into catastrophic decline without great difficulty.[19] There are three elements needed in large quantities by plants: potassium, nitrogen, and phosphorus. Potassium leaches out of granite and similar rocks in enough quantity to be in reasonable supply.

Nitrogen used by a crop like corn can be restored through nitrogen-fixing bacteria in the roots of legumes, alders, and other plants, a process that can be done through crop rotation. Traditional farming systems took advantage of this. Widely, in upland

Southeast Asia, alders are planted after a field is cultivated. The alders grow, fix nitrogen, fertilize the soil, and then become excellent timber when the field is re-cut. In Oceania, *Casuarina* trees (unrelated but also good nitrogen fixers) are used the same way. Native Americans in the northeast allowed leaf litter and wild animal and bird dung to refertilize fields and could practice fairly high-yield continual cultivation of corn for decades.[20] Today, most nitrogen used in agriculture is in the form of fertilizer added to the soil. This fertilizer is produced using the Haber-Bosch process, which uses electricity to convert atmospheric nitrogen (N_2) into ammonia (NH_3). The impact of this method is so great that "about 50 percent of the nitrogen atoms in humans today originate from this single industrial process."[21]

Unfortunately, the other great nutrient needed by all plants, phosphorus, is a very different case. It is widely disseminated in nature but in quantities too small to mine for fertilizer. The main exception is the guano (bird poop, also rich in nitrogen) deposits at seabird rookeries, either current (as in Peru) or fossil (as in Morocco), or sometimes both (as in. Nauru). When these guano deposits are exhausted, the world will be in deep trouble, especially since so much of it depends on these few sources.[22] A bitter irony is that modern fishing and other assaults, including climate change, are rapidly wiping out the seabirds, thus eliminating any future concentration by this method.

An ancient way to obtain guano for fertilizer can be seen in the Middle East. There, one observes large tower-like facilities containing hundreds of places for pigeons and doves to nest (Figure 7.2). The buildings are hollow on the inside, and the guano accumulates on the bottom of the building, where it is collected and spread onto fields. While each structure generates only enough guano for a few fields, there are many of them, providing considerable fertilizer. It is an elegant solution requiring no electricity or chemical input.

Human waste is also a common fertilizer used in many places. This material, often called "night soil," is rich in nutrients and is composted or otherwise processed to kill any organisms within it and put to good use. Even in the US, material from sewage treatment plants is sterilized and sold to farmers, although there is a negative view of this among the public.[23]

Another problem is plant disease; "growers worldwide lose between 10 percent and 23 percent of their crops to fungal disease every year, and another 10–20 percent post-harvest."[24] One way to avoid this issue is the use of organic, low-input, and natural-process farming. While these are widespread and successful, they are small scale and are less productive than the giant farms of agribusiness.[25] This is probably due to a lack of skilled operators at such levels, so it is a solvable problem. It needs to be resolved soon.

Fortunately, there is a good alternative developed: permaculture.[26] This is a modern technology based on Chinese diversified intensive agriculture, as seen in southeast China and Southeast Asia. Agriculture there, based on irrigated rice but growing hundreds of different kinds of plants and animals, is based on maximal use of nutrients before they escape the system. Nutrients trapped by forests in the

FIGURE 7.2   Pigeon towers at the Siwa Oasis, Egypt, in 2016. Photo by Michael Hermann (CC BY-SA 4.0, Wikimedia Commons).

mountains are eventually washed into streams. The water is used to irrigate vegetables and rice, and then to farm fish. Rice paddies grow nitrogen-fixing algae. Ultimately, nutrients escape into the sea to fertilize oyster beds and marine fisheries. No nutrient escapes, and every nutrient molecule is potentially used many times. Permaculture adds more crops from elsewhere in the world and adds modern chemistry and agronomic science, with its detailed knowledge of exactly what is needed by plants and animals. It also allows mechanization on a small, efficient, limited scale. Variants of the permaculture system may offer the best alternative to modern industrial monoculture agriculture.

Some seemingly minor differences are huge. Shade-grown coffee preserves forest ecosystems, with full benefits. Sun-grown coffee on monoculture plantations produces a total desert apart from the coffee plants; virtually nothing else lives in such fields.[27] Drinking shade-grown coffee, especially if certified bird-friendly, is the easiest way for a single person to make a really big difference in the world's environmental problems. Unfortunately, the major corporations and coffee shop chains refuse to play ball. You have to look.

Bee Kind

Another problem is the loss of pollinators.[28] Bees and wasps (Figure 7.3) are only the most visible of these. Squash bees are required to pollinate squash and melon plants. Bumblebees pollinate clover. A huge range of bees, wasps, flies, beetles, other insects, and even birds and mammals pollinate flowers. Hundreds of crops depend on this, from apples and avocados to zucchini. Hand pollination works for some high-value crops, such as dates and apples in China, but we are far better off saving the insects. Massive drenching of the environment with pesticides, especially such deadly killers as neonicotinoids, is exterminating pollinators widely, with potentially catastrophic results.

We suspect that most people would agree that plants are important. They form the basic food for everyone; even carnivores are dependent on them since they eat animals that eat plants. While some plants pollinate themselves, most plants reproduce sexually, with pollen (e.g., plant sperm) being distributed to the female plant parts. Some plants, such as many trees, release their pollen into the wind, which carries it to other trees. The yellow "dust" on your car in the parking lot is wind-distributed pollen, and the allergies you get are often from pollen in the air.

Most of the plant foods we eat come from flowering plants. Most of them rely on insects to distribute their pollen. The USDA estimates that some 75 percent of the world's flowering plants and about 35 percent of the world's food crops are dependent on insect pollinators. Thus, we are dependent on insects; without them, our food supply is endangered, and we do not eat. Pretty straightforward. Grains do not need insect pollination, but most other crops do.

FIGURE 7.3   A wasp covered in pollen on a flower. Photo by Jon Sullivan (Wikimedia Commons, Public Domain).

Bees, butterflies, and similar insects are the primary pollinators, so we are dependent on those insects to grow our food. Farmers, of course, know this; they promote bees as best they can. Some farmers hire commercial beekeepers to bring bees to their fields to pollinate the crops. If you have driven past farms, you may have seen groups of large white "boxes" next to fields; these are the bee colonies brought in by the beekeepers. A nice byproduct of this is the production of honey.

The problem is that bee and other pollinator populations are declining worldwide. While there continues to be some disagreement and continuing research about the cause, with bee parasites being an obvious problem, the major culprit is the pesticides used by farmers to kill the insects that damage their crops. A wide range of pesticides is used for a variety of purposes, and many of those aimed at insects indiscriminately kill insects of all types.

Pesticides are tailored to kill small animals, but given enough exposure, they can also kill or sicken larger animals, including humans. Farmworkers accidentally sprayed with pesticides will get sick, an unfortunate situation in some agricultural areas. The long-term effects of pesticides on humans are unclear.

What can be done? We need pollinators, but we also need to control harmful insects. Developing pesticides that do not kill pollinators would help. Using biological controls, such as releasing ladybugs on plants infested with aphids, is another good step. In China, farmers use ducks to eat the insects in the rice fields, and then eat the ducks. This sort of creative biocontrol needs to be more widespread. More useful in the long run, though, is diversifying farmlands so that vast monocultures do not provide ideal habitats for pests and diseases that spread from field to field. Small, diverse fields separated by hedgerows, meadows, and windbreaks have fewer pest problems and more diversity of pollinators and pest-eating predators.

Increasing bee habitat is also a good step. Some cities have stopped "manicuring" their parks and open spaces so that flowers can grow and encourage bee populations. Replacing your lawn with plants bees love, like lavender and sage, is a very valuable step. In other cases, agencies and even some individuals have set aside land as bee habitat. For example, the actor Morgan Freeman dedicated his 124-acre ranch as a bee sanctuary. Small bee habitats, such as at your home, can also be useful. It would also help not to destroy bothersome bee colonies but to have them removed by beekeepers. Finally, just being aware of the issue is helpful. The Minnesota Board of Water and Soil Resources will pay homeowners to replace their lawns with bee-friendly wildflowers, clover, and native grasses.

People on the Farm

Another problem of large-scale industrial farming is the loss of small farms, and small-scale farmers. Zia Mehrabi points out that the number of farms in the world—about 616 million—has probably peaked. It has declined steadily in North America and Europe for decades, and the process is set in motion elsewhere.

Farms will decline from 2030 or 2040 in other parts of the world, dropping to around 277 million (give or take a few million) by 2100.

This is a real loss. As Zia Mehrabi says, "consolidation brings risks, including loss of indigenous knowledge, loss of connection between society and the land, concentration of corporate power in food systems, a decline in…crop diversity and nutritional diversity in rural landscapes, and lower production in food for direct consumption."[29] Animal welfare may also be affected. Conversely, giant corporate farms are owned by absentee landlords who may care nothing about farming, managed by qualified but often narrow-minded personnel, and worked by highly exploited and underpaid labor in all documented cases. All this guarantees inefficiency, waste, and a lack of environmental concern, as well as injustice and harsh conditions.

Small farmers have the skills necessary to manage mixed farms and less industrial farms, especially in tough conditions. Whether they are old-time small farmers in the US, contemporary Maya farmers in Mexico and Guatemala, Chinese rural people facing the modern world, or any local small-scale or Indigenous farmers anywhere, they are losing their farms to giant corporate interests. Sometimes this is done in the name of "socialism," sometimes it is supposedly "capitalism," but the giant firms are heavily subsidized (and thus de facto socialist). Sometimes it is simply market and cultural forces. Eduardo Brondizio and colleagues report that subsidies to farming worldwide run to $540 billion. This is a very conservative estimate; the real amount is certainly higher.[30]

Redirecting this from large-scale to reasonable-scale farming would help enormously with world food production, especially with efficiency, sustainability, and local self-reliance. Brazil's subsidies for cattle ranching, for instance, should certainly be redirected to more environment-friendly production, including such farming groups as a fruit and spice cooperative, CAMTA.

ENA's father left his natal Texas farm and went to the city, and ENA has since watched many more young people leave similar farms in Texas, China, Mexico, and around the world. Brondizio and colleagues researched small farming and report: "[O]ver the past 30 years, around 200 million jobs in food production have been lost globally…According to our analysis, the current pace could quicken, resulting in at least 130 million more jobs being lost by 2030."[31] The group found that most of the young people they surveyed had a low opinion of their rural situations and wanted to move out.

Yet all research shows that small farmers are much more efficient and can be much more productive than the giant corporate farms beloved of both socialist and capitalist planners. The mindset of government planners and giant-corporation benefactors is often fixated on the alleged benefits of industrial farming because of a completely mistaken and wrong-headed belief that small-scale producers are backward, ignorant, and low-class. Small-scale farmers tend to lack capital and modern inputs, but they have accumulated skills and knowledge of local conditions, diverse plants and animals, dealing with mixed farming (as opposed to monoculture), social and cultural realities, and other matters that make them far superior to corporate farmers, especially in matters of sustainability and efficiency. This is widely studied and realized and reported in detail in several classic works, including Robert Netting's *Smallholders, Householder*, and James Wood's *The Biodemography of Subsistence Farming*.[32] It also appears in many local studies.

Brondizio and colleagues argue for shoring up rural society—providing schools, health care facilities, transportation opportunities, and all the other social goods. These are disappearing all too fast from many rural areas, including ENA's father's east Texas home. They have never reached some areas, including the Maya rainforests, in the first place.

Farming and Disease

All organisms, plants and animals, are hampered by disease, and in industrialized farming, they are placed together in close quarters—situations that almost encourage disease. Dependence on individual species, such as corn grown in vast monoculture farms, is a recipe for disaster. Animals in close quarters are also candidates for infections, and hundreds of millions of animals are killed each year to stop the spread of animal diseases such as swine flu and bird flu. Also of concern is the transmission of animal diseases to humans (zoonotic), such as the bird flu that killed more than 50 million people in 1918 or COVID-19, likely a bat coronavirus. We have the mistaken belief that these diseases are not a US problem. But:

> More zoonotic diseases originated in the U.S. than in any other country during the second half of the 20th century. In 2022, the U.S. processed more than 10 billion livestock, the largest number ever recorded and an increase of 204 million over 2021. Risks occur across the supply chain.... The ongoing H5N1 avian influenza outbreak has left 58 million animals dead in backyard coops and industrial farms. It has infected animals in one of the dozens of live poultry markets in New York City.... Since 2011, the U.S. has recorded more swine-origin influenza infections than any other country. Most occurred at state and county fairs, where an estimated 18 percent of swine have tested positive. These fairs attract 150 million visitors each year... Each year, the U.S. consumes an estimated 1 billion pounds of 'game' (elsewhere called 'bushmeat'). Yet, most hunter-harvested meat is not inspected, and no sanitary measures are required. Avian influenza has spread from wild birds to hunters and also appeared in captive game farms, where 40 million birds are raised annually. .... More than 200 million live wild animals enter the U.S. each year, most undergoing no health and safety checks.[33]

This and other such concerns have led to the One Health movement, committed to considering human and nonhuman animal health as one system. Veterinarians inspired and organized this movement, but it has spread like wildfire among doctors and public health authorities. COVID-19, originally from bats and transmitted to humans by raccoon dogs sold as food (and by careless medical authorities), caused this movement to go from radical fringe to mainstream almost overnight. Not only domestic animals but also wild ones are now considered.[34]

Out to Pasture

About twice as much of the Earth's surface is good for grazing as is good for farming. Animals that can convert grass and brush into human food are thus a part of reasonable use and management of the landscape, vegans to the contrary notwithstanding. The world's best grasslands are usually dominated by perennial grasses that produce enormous root systems, often called "bunchgrasses" in western North America. The grass is only a foot or so high but may have roots going down 10 feet, producing a dense mat of fibers, the sod.

The introduction of steel plows after the US Civil War was the technological innovation that allowed the penetration of this sod for planting, with the farmers being known as "sod busters." The same process occurred in the grassland steppes of Ukraine, Russia, and Kazakhstan. The resulting conversion of prairie grasslands to the production of corn, wheat, and soybeans led to soil erosion and eventually massive loss of nutrients and the emission of considerable carbon. Erosion control and fertilizer use have helped, but without the perennial grasses, soil degradation is inevitable. Tragically, after cultivation or overgrazing, most of the still-grass-covered ranges of the western US have converted to weedy annuals with miserably small root systems. Soil decline continues apace, seriously threatening future food.

To successfully raise domesticated animals in pastures rather than feed lots, one must know a great deal about the animals in the pasture, the availability of water, herd composition and size, movement of herds, and products. Pastoral animals are mostly either grazers or browsers. Grazers primarily eat grasses and low-growing plants, while browsers primarily eat the foliage from bushes and trees. If the animals are browsers, a good pasture would consist of bushes and trees, not grasses.

Grazers generally eat grass but will also do a bit of browsing. Cattle, yaks, sheep, llamas, alpacas, and even horses (more rarely) can live on brush and tree leaves. Most require good pasture and water to do well. Most grazers will eat the blades of the grass but leave the roots in the ground to grow again. Cattle and horses can often share pasture as long as the total number of animals is not high. However, sheep will graze to the roots, and goats may even dig up the roots of the grass and eat them too. Thus, sheep and goats are more likely to overgraze and do not share pasture with other grazing animals well.

Browsers generally eat the foliage of bushes and trees and include goats, camels, and reindeer. Goats can both browse and graze, making the species very versatile and adaptable to most pasture types. Goats will eat most vegetation and can seriously damage or even destroy pastures. The desert-looking areas of much of the Middle East (e.g., Palestine, Israel, Jordan, Lebanon, and Syria) were much greener before goats ate much of the vegetation and firewood collection took much of the rest.

Camels can get along well on poor pasture and with relatively little water. Both water and fat are stored in the hump(s), and camels can survive in severe conditions for long periods of time. Reindeer browse on the short vegetation of the tundra, primarily lichens, and are herded by various traditional groups in northern Scandinavia and Russia. Pasture for reindeer is available all year, and the animals move around the landscape on their own to take advantage of better areas.

Properly managed, pasture lands can produce many animals for meat, milk, leather, and other goods without competing with farmed crops. An example of the management of pastures for cattle are the Maasai herders living in southern Kenya and northern Tanzania in eastern Africa. The Maasai occupy a region with two major ecozones: a relatively arid plain with extensive grassland and a better-watered upland area mostly occupied by farmers. To the Maasai, cattle form the basis of life.

Cattle are critical resources and are used as currency, to legitimize marriages, to solidify social relationships, and as status symbols. Cattle also provide food and materials. Cattle are not butchered for their meat (a one-time use) but are exploited for their milk and blood, resources that can be used over time. Blood can be taken from a cow about once a month (but as there are many cows, blood is always available), mixed with milk, and consumed as food. The dung from the cows is used as fuel and as plaster in Maasai houses. When a cow dies, the meat may be eaten and the other parts of the cow (e.g., hide and bone) are used.

Pasture is also a critical resource, and animals must be moved often. There is a complex system of who can have how many animals on which pasture, and this is controlled by the local leaders. Pasture conditions must be constantly monitored, and animals are closely watched as to their general health and ability to produce milk to determine the quality of the pastures. Without good pasture management, the cattle could not thrive.

Unfortunately, the Maasai are under extreme pressure from a general drought in the region (the result of climate change), and they have had to drastically reduce the number of cattle in their deteriorating pastures. In addition, farmers are trying to take over Maasai lands, and conservation groups are trying to set aside Maasai pasture lands for protected wildlife habitats.

Tension in other African areas has escalated to violent levels as climate change dries up the range. Conflict is avoided where open grasslands are abundant enough that everyone can pasture their livestock, but fewer and fewer areas are so lucky.[35]

Herding is at least as well managed in Mongolia as elsewhere. Traditionally, management was so careful that there was *more* grass around waterholes rather than less. In most rangelands, heavy use of waterholes scuffs away grass.[36] However, the pattern of privatization, fencing, and rapid collapse of the range is being repeated there. This is true especially in Inner Mongolia, which is part of China. The Chinese fence the range, expand cultivation into marginal areas, and otherwise damage range management, with often disastrous consequences; desertification is quite visible on Google Maps satellite photos. As in East Africa, privatizing—or at least breaking up and fencing—the range leads to devastation, where commons management was successful for thousands of years (ENA has observed this process in both places).

Mongolia continues to allow free-range, usually common-pool, management, with success in managing the rangelands. Far from the nearest trail (let alone road), ENA has walked with Mongolian herders who could tell the use to animals and humans of every tiny plant. Traditional ideology of the high latitude grazing societies is notably conservationist, stressing respect (*shuteekh* is one Mongol word).[37] However, urbanization has led to a concentration of herders in the periurban zone around the capital, Ulaan Baatar. People move to the city for opportunity but keep their *ger* houses (the "yurts" of popular usage, from a Turkic word for "home place") and a few animals. The result is that overgrazing has destroyed the grass in a widening zone around the city.

Common in Europe, western North America, and western Asia is *transhumance*: moving animals from winter pasture to summer pasture along familiar routes. Relatively local herding and short-distance travel can also be well managed. The Swiss have also managed transhumance for millennia without degrading pastures. Strongly held ownership and careful restriction of pressure are key in that case. Stépanoff and colleagues compare several forms of herding, from long-distance migration to caring for animals on farms and also hunting, in modern Siberia.[38] As usual, groups that depend on stock raising know their range manage well. They use respect and even religious veneration of their animals to do it, as do East African groups.

Mismanagement of pasture lands and overgrazing in the US was the result of an explosive growth in cattle ranching beginning in the late 1800s, where the animals were grazed without care for the grass resource. This devastated much of the US rangeland with incredible speed. One of the problems is that much of the pasture lands in the western US are owned by the government and leased to ranchers. It costs the government about seven dollars per acre to maintain the range, but the ranchers pay (some refuse to pay even that) as little as three dollars per acre to lease it. Thus, the ranchers pay little. Many overgraze and trash the land, and then want the government to pay to fix it. Another issue is the government paying to level forests to create pasture to lease to ranchers at a loss. Makes no sense.

Some of the worst damage was in southeast Arizona, whose enormously rich grasslands had turned to bare powdery ground by the mid-1890s. This finding started a debate that still goes on. It turned out that drought in the 1890s after previous wet years was part of the problem. Another was beaver trapping in the 1830s, which eliminated beavers from southern Arizona. Beaver dams had held back the rivers and streams, producing chains of lakes and ponds. Without those dams, the streams promptly cut down into the soil, forming deep gullies. The water table, previously kept high by beaver dams that let water pond up and percolate into the soil, dropped catastrophically. Springs and water holes went dry, creeks ran only after rain, and even rivers became seasonal. Erosion stripped off the best soil. Into this came the cattle, whose thousands of hooves further hurt the soil and made it erode or compact into rocklike adobe.

Moreover, fire control and elimination of grass led to invasion by woody leguminous plants, especially mesquite and acacia. They permanently converted vast areas from grassland into short thorn-tree woodland. The lowering of the water table greatly helped this process, since the leguminous trees can send their roots down deep while the grass is not so deep-rooted.

In the 1890s, the cattle industry in Arizona collapsed and thousands of cattle died. The range and Arizona's cattle industry recovered slowly, with lower stocking rates and regrowing grass. Another drought in the 1950s caused further trouble. Much of the story is told in *The Changing Mile* by James Hastings and *The Changing Mile Revisited* by Raymond Turner and others.[39]

Damaged grasslands can be repaired. The grasslands of the Altar Valley in Arizona, a desert that requires the most careful management, were restored by local effort.[40] Similar efforts in Nevada have been effective at protecting range, but most ranchers in that state have overused government lands and allowed grazing and fire to degrade land massively, endangering native species of birds, plants, and mammals as well as the cattle industry. Self-consciously conservation-oriented ranchers have shown the range can be saved, but it takes effort and will.

There has been a great deal of destruction of forests to create pastures. This began long ago; much of Europe was cleared for livestock by 5,000 years ago. Middle Eastern forests slowly gave way also, frequently with wild fruit trees remaining to show that what is now a pasture was once a woodland. Today, rainforest clearing is rampant (Figure 7.4). Pinyon-juniper woodlands in the western US are also massively eliminated to make pasture, though erosion soon destroys the soil, and the value of cattle produced is low. Junipers can be invasive and weedy, but pinyons (nut-bearing pines) are essential to wildlife and are very slow-growing, so this is an ecological disaster.

Bugging Out

A food of the past, present, and perhaps future is insects.[41] Most societies, including most contemporary ones, routinely include insects in their diets. And, like it or not, insects (and some other things such as rodent hair and feces) are in all the processed foods eaten by Westerners. In addition to food for people, insects make excellent animal feed, freeing up farmland used to grow food for animals to instead produce food for people.

So, what's the problem? Insects are an excellent source of protein, abundant, easy to obtain and/or raise, and only a few have any detrimental effects. The issue seems to be one of classification and associated repugnancy and phobia. Western societies typically classify insects as pests, dirty vermin to be killed. We spend a great deal of effort and poison to kill them.

FIGURE 7.4   A rainforest in Borneo being deforested. (iStock image 1257301516).

Other societies put insects into different classificatory categories, some as pests to be sure, but others as food, medicine, and technology (e.g., silk). Insects commonly used as food include grasshoppers, caterpillars, and various larvae. These, and many more, can be seen for sale in traditional markets across the planet, and a few are available in Western stores as well. There is nothing wrong with them.

In ancient Lake Texcoco, now occupied by Mexico City, one of the animals living in the lake was a shore fly (*Ephydra hians*). The flies laid their eggs in the water, and the pupae floated to the surface where they would hatch into flies. Thus, the surface of the lake had extensive mats of shore fly eggs, pupae, and adults. The Aztec gathered this shore fly material as food. This resource was so important that the genus was once renamed *Hydropyrus*, Greek for "water wheat" (but now known as *Ephydra* again).[42]

Protein Overload

We all need protein in our diet, but protein is not a single thing. Protein includes at least 22 different amino acids. Some can be manufactured by the body, but eight (called essential amino acids) must be obtained from food. When foods are eaten, the proteins within them are broken down into their constituent amino acids, which are then used as building blocks for new proteins. Although all foods contain protein, most do not contain all 22 amino acids. If one does not get enough of the essential amino acids, the body will begin to break down its existing protein, essentially digesting itself. In the extreme, this can result in death.

Protein is present in plants but is found in higher concentrations in animals. The kicker is that one of the essential amino acids, lysine, is found primarily in animal proteins. Methionine is also rather sparse in plant foods. While it is possible to be a strict vegetarian, it is necessary for the person to have extensive knowledge about which combinations of plants to eat, in what sequence, and with what frequency. The best combination is grains and beans, and all traditional farming societies have figured out grain-and-bean dishes to allow survival. Even then, one still needs vitamin B_{12} in the diet, and that must be obtained either from animal or fungal sources or from artificial supplements. Most people lack such knowledge; they simply consume some animal or fungal protein to meet their overall needs.

Along these lines, the ancient Mesoamerican diet consisted of the "three sisters," corn, beans, and squash—a diet that provided the essential amino acids. Chile also aided this diet, supplying vitamins. Lysine, although present in beans and some other plants, was most easily obtained by eating some animal products such as milk, eggs, or meat. On the other end of the spectrum is the traditional Inuit diet, one that consisted primarily of animal protein (e.g., seals). In a diet high in meat, it is necessary to have sufficient fat and/or carbohydrates; an issue the Inuit solved by eating all of the animal; meat, fat, organs, and stomach contents, all of it. Interestingly, one can "starve" to death by eating only lean meat since it lacks the essential fats and carbohydrates.

In reality, only a small amount of animal protein is needed to have complete nutrition. People in Western societies consume vastly more protein than is required by the body in the typical "meat and potatoes" diet. While the extra protein does not hurt, the fats in the meat can cause some health issues such as obesity and heart problems and as such is not the healthiest diet. People need some ~60 mg/kg of lysine in infancy to ~30 mg/kg in adults, and most people today get their lysine by eating some animal products.[43] An egg a day would do.

The other issue with eating so much meat is that meat production requires enormous amounts of energy, feed, and water. These resources could be much better used elsewhere.

The Beef with Beef

There are many species of wild and domesticated cattle, some of which are now extinct. Domesticated cattle number some 1.3 billion and are found in most regions of the world. Some cattle are specialized, such as beef (meat) or dairy (milk and cheese) cattle, and some are used for labor. Beef cattle are raised specifically for their meat, with steers (castrated males) being the primary animal utilized. In 2022, some 28 percent of the beef consumed worldwide originated in the US, while other major contributors include Brazil, China, and the EU.

The average steer will produce some 425 pounds of meat, with perhaps 200 pounds being made into hamburgers with the remainder being cut into steaks, ribs, roasts, and the like. The remainder of the animal is used to make a variety of other products, such as leather, glue, pharmaceuticals, cosmetics, soap, and pet food.

Some cattle are raised on pastures, eating primarily grass, but in lean years or in bad weather, they will be fed alfalfa. In the US, most beef cattle are raised or finished in feedlots, where they are fattened up for slaughter. The cattle are fed various grains and other materials, such as molasses, field corn, and soybeans. Beef contains considerable protein plus a variety of other nutrients.

In 2020, some 130 billion pounds of beef (something like 300 million cows) were consumed worldwide, with Americans eating some 28 billion pounds. The next highest consumers (in total national terms, not per capita) were China, the EU, Brazil, and India. The US accounted for roughly 21 percent of the beef consumed in the world in 2020. Americans seemingly cannot live without their hamburgers, steaks, ribs, and hot dogs.

So, what's the beef with beef? The issues are threefold. First, raising beef cattle is a highly inefficient use of resources. It takes about 10 calories of feed to make one calorie of cow (a 10:1 conversion ratio). Cattle are fed field corn grown for animal feed on farmland that could be used to grow sweet corn that could directly feed people. While there are some nutritive differences between field and sweet corn, the differences are minor.

A typical can of sweet corn from the store has about 350 calories, while a pound of average hamburger has about 1,500 calories. Simple math using the 10:1 conversion ratio shows that it takes 15,000 calories (more than 43 cans of corn) to make a pound of hamburger. Those 15,000 calories of corn are enough to feed seven or eight people, while a pound of hamburger seemingly cannot feed even one. Thus, feeding "people food" to cattle is very inefficient (although some cattle eat grass that people cannot eat, making the pasture method fairly efficient). For comparison, pigs have a 5:1 conversion ratio, and commercially grown chickens have about a 2:1 ratio (they do not move around much). And this does not even include the massive amount of water needed for cattle.

Second, while cattle can eat alfalfa, people cannot (except as sprouts). Thus, growing supplemental cattle food such as alfalfa removes that agricultural land from the production of food that people can eat. Alfalfa demands a great deal of water and sun, so it is usually grown in deserts and near deserts, making it one of the biggest reasons for groundwater depletion and river loss worldwide. Again, very inefficient.

Third, the production of beef cattle in rainforests such as in Brazil is done using pastures. Natural pastures are generally good, but to make pastures for cattle in forests, the forests are cut down and destroyed (now using bulldozers and chainsaws, and not only in Brazil) and grass planted. These new pastures are productive for a few years, but then the soil becomes exhausted (forest soils tend to be poor to begin with), and a new pasture must be made by destroying another chunk of forest. The old pasture is abandoned but can rarely recover since it is surrounded by other abandoned pasture fields.

Think of the forest as your skin. If you scrape yourself, it might hurt, but there is very little real damage, and your skin will quickly heal. If you are seriously burned and lose, say, 25 percent of your skin, you will be in the hospital and probably need skin grafts, but you should survive. If you lose 75 percent of your skin, you will likely die since there is little prospect that the remaining 25 percent of your skin can "cover" you. The forest is like skin since the more forest that is lost, the greater the likelihood that it cannot recover and will die. Then we are all in trouble.

So, what is the true cost of the average hamburger? First, of course, is the money spent at the grocery store or restaurant. Then calculate how many other people will go hungry so you can have a BBQ. Then figure the average loss of forests plus all the lives of the many forest plants and animals killed during the destruction of the forest to make a pasture. Finally, figure in the methane cattle produce.

Adding insult to injury, Americans use about a million tons of briquettes every year in their BBQs. Most of the charcoal briquettes used to BBQ all this beef are made from charred wood byproducts and coal, along with various additives to make them burn consistently. As with hot dogs, you probably do not want to really know what is in them. Other briquettes are made from charcoal alone. In each case, hardwood trees are cut down, fragmented, burned, and processed into the familiar briquette form. We then use lighter fluid (made from oil) to reignite them to cook our hamburgers. The loss of trees in this process is bad enough, but many of the charcoal briquettes used in American BBQs come from rainforest trees, contributing to the degradation and destruction of those forests.

We are not advocating that people should never eat beef. We are suggesting that we all eat it only when sustainably produced in natural pastures. Such beef production is common today, using natural grasslands that are not overgrazed. Rotation systems allow heavy stocking without degradation, as we know from African herder experience. Locally tailored systems with native grasses have been developed from Arizona to Oregon and from New England to China. In addition, eating less beef means more food for other people, a reduction in the loss of forest, greater preservation of biodiversity, and of the preservation of the lives of the creatures of the forest that would otherwise be lost.

There are now substitutes to real meat in the form of plant-based meat substitutes. As the quality, taste, and texture of these products improve, they will eventually become a reasonable alternative to animal meats. In addition, real animal flesh can now be grown in the lab without having to raise and slaughter a complete animal.

The Past and Future of Farming

Here we describe several ancient systems that were sustainable for many centuries. Each ultimately failed—one to drought and invasion, one to government action, and one to population loss from introduced disease. Still, each can teach us about how we may adapt our farming to a more sustainable place, all while having less impact on the environment. A win-win situation if we listen and act.

Ancient Maya Agriculture

The ancient Maya lived in southern Mesoamerica and employed a very complex agricultural system that was able to support huge populations in large cities within the region's rainforests. Their descendants, the contemporary Maya, still live there, still speak Maya, and still use the same agricultural system, albeit rather modernized. This system was sustained for more than a thousand years, until drought and possibly overpopulation overtook it about 900 years ago. Still, some of the Maya cities persisted for hundreds of years longer. The last Maya city was conquered by the Spaniards in 1697.

Early on, archaeologists investigating the Maya discovered large buildings in the rainforest. The large buildings, generally arranged around a central plaza, were all that could be seen in the thick rainforest. Accordingly, these sites were interpreted as ceremonial centers for a dispersed farming population. Archaeologists eventually noticed low mounds of soil and, upon investigation, discovered they were the foundations of houses made from wood that had long since decayed. Once they started looking for these subtle features, they were found everywhere. It was realized that the sites were not ceremonial centers but cities with large populations (Figure 7.5). Most recently, the new method of LiDAR (Light Detection and Ranging, essentially the laser equivalent of radar) has revealed even larger and more complex cities, fields, irrigation systems, roads, fortifications, and the like. Truly remarkable.

How were the ancient Maya able to feed so many people? They had no machines and no draft animals, just human labor. Archaeologists began by looking at the contemporary Maya farming system for clues. They found that the contemporary Maya used many types of small gardens, strategically placed trees (such as ramón trees for nuts and beauty), and a fantastically detailed knowledge of the region's plants and animals. The small gardens included chinampas, terraced fields, slash-and-burn fields, and orchards. They also utilized wild plants and animals.

Chinampas are small, raised fields constructed in swampy areas. The Maya would dig ditches in wetlands and pile up the soil in a waffle-like pattern, raising the fields out of the water and creating canals between them. Soil dredged from the canals was added to the fields when necessary to maintain their fertility. The canals were colonized by turtles and fish, both of which the Maya ate. The Mexica (including the Aztecs) also built chinampas on Lake Texcoco around their capital city of Tenochtitlán (now under Mexico City), and a remnant of that system, the Floating Gardens of Xochimilco (Figure 7.6), is a tourist attraction in Mexico City today. On the hillsides, the Maya built small stone-walled enclosures and small terraced gardens.

In the forest proper, the Maya use the swidden technique where a small field is selectively cut, the vegetation burned, and the ash spread out on the field. This slash-and-burn field could grow a crop or two, then had to be abandoned due to its poor soil. With careful management, the same field could be used again in

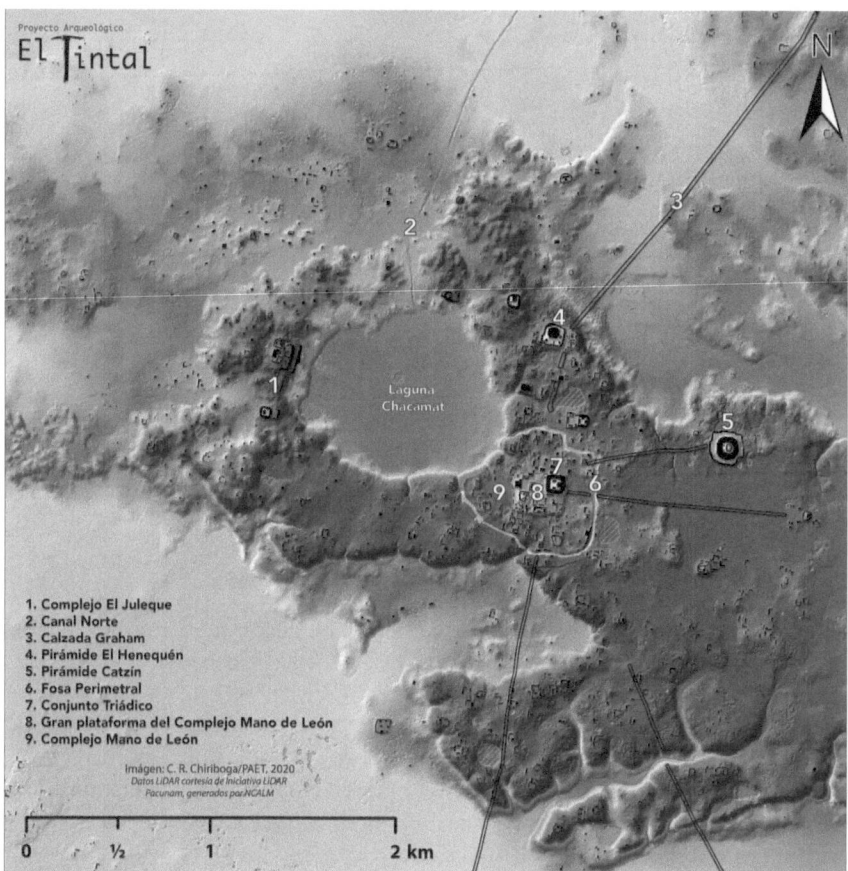

Proyecto Arqueológico
El Tintal

1. Complejo El Juleque
2. Canal Norte
3. Calzada Graham
4. Pirámide El Henequén
5. Pirámide Catzín
6. Fosa Perimetral
7. Conjunto Triádico
8. Gran plataforma del Complejo Mano de León
9. Complejo Mano de León

Imágen: C. R. Chiriboga/PAET, 2020
Datos LiDAR cortesía de Iniciativa LiDAR
Pacunam, generados por NCALM

FIGURE 7.5 Map of the Maya archaeological site of El Tintal, generated with the LiDAR technology. Photo by Carlos R. Chiriboga/PAET (Wikimedia Commons, Public Domain).

10–20 years (or, at extremes, 5–50), making swidden sustainable. The regrowth, which took up enough carbon to make up for the original burning, was often managed to maximize nitrogen-fixing plants that fertilize the soil. Orchards were placed where they would grow best or in places too rocky to otherwise farm. Regrowth was managed such that useful firewood trees and nitrogen-fixing leguminous trees flourished; useless trees were often weeded out. Dooryard gardens were also important, holding up to about 100 species of plants, as well as a few domestic animals (turkeys and dogs were the only significant ones). While these farming methods generate GHGs, they are far less than industrialized methods.

Using these clues, archaeologists began to look for the remains of these types of fields, and once knowing what they were looking for, they were found everywhere. Thousands upon thousands of them. They also found complex irrigation

FIGURE 7.6   A chinampas note the posts to stabilize the fields in the canals. Photo by Emmanuel Eslava (CC BY-SA 4.0, Wikimedia Commons).

canals and reservoirs and an extensive road system for the transport of goods. The ancient Maya used the same basic methods as today, but in a much more intensive manner. Gardens, fields, and orchards were placed everywhere, as determined by local conditions. The rainforest was heavily modified but not destroyed as in the contemporary Western system. Together, these methods provided a productive and sustainable agricultural system that supported the Maya for many centuries.

How does this apply to us? The Western approach to farming in the rainforest is to level it, plant a few crops, graze a few cows, exhaust the soil, and then move on to the next chunk of forest, leaving behind a devastated landscape. But an understanding, a reverse-engineering of sorts, of Maya practices could lead to the development of techniques that could be utilized in contemporary rainforest settings, perhaps supporting large populations without the destruction of the forest. If the Maya system could be adapted and adopted, rainforests could become very productive places capable of supporting large numbers of people in a sustained manner without sacrificing the forest. Some Maya are working on such systems today.[44]

The Traditional Chinese Agricultural System

The basic agricultural system of southeastern China relied on rice as the staple food, providing up to 90 percent of calories. It was produced in irrigated paddy fields. The water came down from the hills and mountains that make up most of the south Chinese landscape. Mountain forests were originally tall and lush, but increasingly over the years they were burned for shifting cultivation or to prevent bandits and tigers from hiding in them. The forests and brush regrew rapidly, often with nitrogen-fixing plants, and plants with root systems that could tap into and mobilize deep mineral nutrients. The liberated ash and soil washed down the streams, steadily increasing the fertility of the system. As long as the burning was not too frequent, the forests withstood this treatment, but eventually they were the weak points of the system, a situation now being remedied by reforestation.[45]

The water was first diverted into terraced vegetable fields around the villages. Houses were situated where streams came together, assuring a water supply, and were carefully kept off the irrigated rice land—far too valuable to be lost to pavement. This was all managed by the site planning art of *feng shui*, "wind and water." It also kept groves around villages and taught people to place villages on the lee sides of hills—protected from wind and water. Modern thinking dismisses *feng shui*, builds houses on good land and in flood zones, cuts down groves, wastes water, and continually demonstrates the old ways were better.

Below the vegetable fields are the rice paddies, flooded in spring and dried off in fall for harvest. They produced not only rice but also fish, frogs, birds, and other edible wildlife. Below them, where water ponds are too deep for rice, there are ponds for water plants and for fish rearing. Eventually the water reaches the sea, and the nutrients in it nourish oyster beds and a huge fishery. Today, with pollution and overfishing, the oyster beds are few and the fish are almost entirely gone. It is important to emphasize that the old system lasted for thousands of years, increasing the fertility and area of the paddies all the time. The current system of industrial agriculture, high chemical use, polluted water, and paving over the best rice land is already collapsing. Young people are abandoning the farms and moving to the cities.[46]

The Ancient Amazonian Agricultural System

Early Europeans exploring the Amazon River reported seeing large and complex societies living along the river and its tributaries. Later explorers found very few people where those societies had first been reported, so the idea of complex societies along the Amazon was summarily dismissed. Many people had seen dark soils in the rainforest, called "dark earth" (*terra preta* in Portuguese), but these were just seen as curiosities. Then, beginning in the 1990s, archaeologists began to investigate the dark earths and found that they were agricultural soils. Archaeologists also found the remains of small cities with earthen architecture, and more urban centers continue to be discovered. It seems that the early explorers were right. There were large and complex societies along the river but had probably been destroyed by disease left by the first European explorers and were gone by the time the next Europeans got there.[47]

How did the ancient Amazonians farm in the poor soils of the rainforest and support such complex societies? The black earth was formed by mulching trash, broken pottery, plant and animal remains, and charcoal (from the burning of cleared vegetation but not burned to ash) and mixing it with the native soils. The new soil was then placed in extensive complexes of raised fields. Though initially labor-intensive to build, once completed, the raised fields were easy to maintain. They could be easily fertilized by adding new mulch and spreading it over the plot.

An understanding of this ancient system is important for two reasons. First, such a system of fertile plots in the rainforest may have direct and practical application, as it may be possible for farmers to successfully farm the rainforest without destroying it. Second, it is possible that the addition of charcoal, rather than ash, into the soil could result in the storage of carbon and a lowering of carbon emissions.

Again, tree crops were vital to the system. The Amazonian forests today are heavily modified, at least where the population was formerly dense. Fruit and other useful trees are more common than in undisturbed areas.

Some Solutions

Farming needs to move away from the huge enterprises that depend on fossil fuels, chemical fertilizers, herbicides, and pesticides, all of which poison the Earth. Even if the fuels and chemicals do not run out, water will. There are solutions, but we must break the cycle we are in and adopt new methods.

Raising and eating fewer animals would help by saving considerable water, using feed that people can eat to feed people, and converting fields from growing fodder to growing food for people. This simple change would increase human food supply dramatically.

Another possibility is to feed animals foods humans will not eat, such as insects. Insects can be raised on an industrial scale, fed on food waste that would otherwise end up in a landfill, then made into animal feed. The insect frass (poop) can be collected and used as fertilizer. Not only is this possible, but it is also already being done with some companies producing hundreds of thousands of tons of animal feed and fertilizer. Of course, humans could (and many do) eat the insects directly, saving the middleman.

In forested areas such as Mesoamerica, the Amazon, Africa, and Southeast Asia, ancient agricultural systems could be reintroduced. Such systems have been shown to be productive within forests without destroying them. Think outside the box.

Individuals and neighborhoods can produce foods locally in small gardens. This can be done in urban areas with community organizations cultivating vacant lots. Fruit trees and small gardens can be placed in backyards. Some food plants can be grown on vertical surfaces, such as the side of buildings. This is reminiscent of the "victory gardens" of World War II that helped food production for the war effort.

Sheep, Goats, Ducks, and Ladybugs

Western-style farming relies heavily on herbicides to kill unwanted vegetation (weeds) that take nutrients and water away from the plants being purposefully grown. Many herbicides are selective about what general kinds of plants they kill (e.g., broadleaf or narrowleaf) and so can be generally tailored to the crop. Weeds in fields can also be removed mechanically or by hand.

However, in some instances, weeds or other unwanted vegetation can be controlled by using grazing animals such as sheep or goats (recall that goats will eat most any plant) instead of poisons. These grazers/browsers can be put on fallow ground to eat the weeds. They can also be placed in orchards and in other locations where plant cover needs to be otherwise managed, such as road shoulders or firebreaks. As a bonus, the animals will also leave natural fertilizer.

Other animals can be used in lieu of pesticides, a biological rather than chemical control. The traditional Chinese system uses herds of ducks, rented to the farmer by the duck herder, to waddle through the rice fields and eat the insects on the plants, leaving fertilizer in their wake. Eventually, after the duck converts the insects into duck, the duck will be eaten.

In ornamental gardens, aphid (and other insect) infestations are always an issue. Poisons will work, but a simpler and more efficient way of aphid control is the use of ladybugs, voracious carnivores that make short work of the aphids. Buy cartons of ladybugs at your garden store.

Notes

1 Vandermeer, John H. 2011. *The Ecology of Agroecosystems*. Sudbury, MA: Jones and Bartlett.
2 Zeder, Melinda. 2012. "The Domestication of Animals." *Journal of Anthropological Research* 68:161–190; Zeder, Melinda. 2015. "Core Questions in Domestication." *Proceedings of the National Academy of Sciences* 112:3191–3198; Isett, Christopher, and Stephen Miller. 2018. *A Social History of Agriculture from the Origins to the Current Crisis*. Lanham, MD: Rowman and Littlefield; ArchaeoGLOBE Project. 2019. "Archaeological Assessment Reveals Earth's Early Transformation through Land Use." *Science* 365:897–902.
3 National Academy of Sciences. 1996. *Lost Crops of Africa. Vol. I: Grains*. Washington, DC: National Academy of Sciences Press; National Academy of Sciences. 2006. *Lost Crops of Africa. Vol. II: Vegetables*. Washington, DC: National Academy of Sciences Press; Van Zonneveld, Maarten, et al. 2023. "Forgotten Food Crops in Sub-Saharan Africa for Healthy Diets in a Changing Climate." *Proceedings of the National Academy of Sciences* 120:e2205794120.
4 Smil, Vaclav. 2013. *Harvesting the Biosphere: What We Have Taken from Nature*. Cambridge: MIT Press, p. 179. See also Harrell, Stevan. 2023. *An Ecological History of Modern China*. Seattle: University of Washington Press.
5 Bren d'Amour, Christopher, Femke Reitsma, Giovanni Baiocchi, Stephen Bartel, Burak Güneralp, Karl-Heinz Erb, Helmut Haberl, Felix Creutzig, and Karen C. Seto. 2017. "Future Urban Land Expansion and Implications for Global Croplands." *Proceedings of the National Academy of Sciences* 114:8939–8944.

6 Montgomery, David R. 2012. *Dirt: The Erosion of Civilizations* (New ed.). Berkeley: University of California Press.

7 Thaler, Evan A., Isaac J. Larsen, and Qian Yu. 2021. "The Extent of Soil Loss across the U.S. Corn Belt." *Proceedings of the National Academy of Sciences* 118:e1922375118.

8 Torres, Aurora, Jodi Brandt, Kristen Lear, and Jianguo Liu. 2017. "A Looming Tragedy of the Sand Commons." *Science* 357:970–971.

9 Figures from David Pimentel's classic research (Counting Calories in Agriculture | MEPartnership, 2006). Some estimates run far higher, especially for US agriculture.

10 Anderson, E. N. 2014. *Caring for Place*. New York: Routledge.

11 Chapron, Guillaume, Yaffa Epstein, and José Vicente López-Bao. 2019. "A Rights Revolution for Nature." *Science* 363:1392–1393.

12 Vandermeer, John H. 2011. *The Ecology of Agroecosystems*. Sudbury, MA: Jones and Bartlett.; West, Paige. 2016. *Dispossession and the Environment: Rhetoric and Inequality in Papua New Guinea*. New York: Columbia University Press; Zanotti, Laura. 2016. *Radical Territories in the Brazilian Amazon: The Kayapo's Fight for Just Livelihoods*. Tucson: University of Arizona Press.

13 Mehrabi, Zia. 2023. "Likely Decline in the Number of Farms Globally by the Middle of the Century." *Nature Sustainability* 6:949–954.

14 Alfred, Brady W., et al. 2015. "Ecosystem Services Lost to Oil and Gas in North America." *Science* 348:401–402.

15 Associated Press. 2023. "258 Million Had Dangerous Food Insecurity Last Year, Report Says." *Los Angeles Times*, May 8, A2.

16 Ben-Ghiat, Ruth. 2020. *Strongmen: Mussolini to the Present*. New York: W. W. Norton; Dikötter, Frank. 2019. *How to Be a Dictator: The Cult of Personality in the Twentieth Century*. New York: Bloomsbury.

17 Hong, Chaopeng, Jennifer A. Burney, Julia Pongratz, Julia E. M. S. Nabel, Nathaniel D. Mueller, Robert B. Jackson, and Steven J. Davis. 2021. "Global and Regional Drivers of Land-Use Emissions in 1961–2017." *Nature* 589:554–561.

18 Renard, Delphine, and David Tilman. 2019. "National Food Production Stabilized by Crop Diversity." *Nature* 571:257–260; Salaman, Redcliffe. 1949. *The History and Social Influence of the Potato*. Cambridge: Cambridge University Press.

19 Lehmann, Johannes, and Markus Kleber. 2015. "The Contentious Nature of Soil Organic Matter." *Nature* 528:60–68.

20 Hurt, John P. O., and Susan Winchell-Sweeney. 2023. "Resetting Archaeological Interpretations of Precontact Indigenous Agriculture: Maize Isotopic Evidence from Three Ancestral Mohawk Iroquoian Villages." *American Antiquity* 88:497–512.

21 Chen, Jingguang G., Richard M. Crooks, Lance C. Seefeldt, Kara L. Bren, R. Morris Bullock, et al. 2018. "Beyond Fossil Fuel-driven Nitrogen Transformations." *Science* 360:873.

22 Barbieri, Pietro, Graham K. MacDonald, Antoine Bernard de Raymond, and Thomas Nesme. 2022. "Food System Resilience to Phosphorus Shortage on a Telecoupled Planet." *Nature Sustainability* 5:114–122; Egan, Dan. 2023. *The Devil's Element: Phosphorus and a World Out of Balance*. New York: W. W. Norton.

23 Owens, Brian. 2023. "Human Waste Could Help Tackle a Global Shortage of Fertiliser" (Ed. This is the spelling used). *New Scientist*, 19 January 2023.

24 Stukenbrock, Eva, and Sarah Gurr. 2023. "Address the Growing Urgency of Fungal Disease in Crops." *Nature* 617:31–34, p. 31.

25 Kravchenko, Alexandra N., Sieglinde S. Snapp, and G. Philip Robertson. 2017. "Field-Scale Experiments Reveal Persistent Yield Gaps in Low-input and Organic Cropping Systems." *Proceedings of the National Academy of Sciences* 114:926–931.

26 Holmgren, David. 2017. *Permaculture: Principles and Pathways beyond Sustainability* (revised ed.). Melliodora, Seymour, Victoria, Australia; Bane, Peter. 2012. *The Permaculture Handbook: Garden Farming for Town and Country*. Vancouver, BC: New

Society Publishers; Mollison, Bill. 1988. *Permaculture: A Designers' Manual.* Sisters Creek, Tasmania: Tagari Publications-1997. *Introduction to Permaculture.* Berkeley, CA: Ten Speed Press.

27 Perfecto, Yvette, M. Estelí Jiménez Soto, and John Vandermeer. 2019a. "Coffee Landscapes Shaping the Anthropocene: Forced Simplification on a Complex Agroecological Landscape." *Current Anthropology*, supplement 20; Perfecto, Yvette, M. Estelí Jiménez Soto, and John Vandermeer. 2019b. "Patchy Anthropocene: Frenzies and Afterlives of Violent Simplifications." *Current Anthropology*, supplement 20:S136–S250. Also ENA, personal research.

28 Potts, Simon G., et al. 2016. "Safeguarding Pollinators and Their Values to Human Well-being." *Nature* 540:220–229.

29 Mehrabi, Zia. 2023. "Likely Decline in the Number of Farms Globally by the Middle of the Century." *Nature Sustainability* 6:949.

30 Brondizio, Eduardo S. Stacey A. Giroux, Julia C. D. Valiant, Jorndan Blekking, Stephanie Dickinson, and Beate Henschel. 2023. "Change Mindsets to Stop Millions of Food-Production Jobs from Disappearing." *Nature* 620:33–36; Netting, Robert M. 1993. *Smallholders, Householders: Farm Families and the Ecology of Intensive, Sustainable Agriculture.* Stanford: Stanford University Press; Wood, James W. 2020. *The Biodemography of Subsistence Farming.* Cambridge: Cambridge University Press; For the Maya, see Anderson, E. N. 2005. *The Political Ecology of a Yucatec Maya Community.* Tucson: University of Arizona Press; For China, see Santos, Gonçalo. 2021. *Chinese Village Life Today: Building Families in an Age of Transition.* Seattle: University of Washington Press.

31 Brondizio, Eduardo S., Stacey A. Giroux, Julia C. D. Valiant, Jorndan Blekking, Stephanie Dickinson, and Beate Henschel. 2023. "Change Mindsets to Stop Millions of Food-Production Jobs from Disappearing." *Nature* 620:33; for the co-op noted below, see pp. 35–36.

32 Netting, Robert M. 1993. *Smallholders, Householders: Farm Families and the Ecology of Intensive, Sustainable Agriculture.* Stanford: Stanford University Press; Wood, James W. 2020. *The Biodemography of Subsistence Farming.* Cambridge: Cambridge University Press

33 Linder, Ann, and Dale Jamieson. 2023. "Blind Spots in Biodefense." *Science* 379:621.

34 Braverman, Irus (ed.). 2023. *More-than-One Health: Humans, Animals, and the Environment Post-Covid.* London: Routledge.

35 On the African herding situation, see Dyson-Hudson, Neville. 1966. *Karimojong Politics.* Oxford: Oxford University Press; Evans-Pritchard, E. E. 1940. *The Nuer: A Description of the Modes of Livelihood and Political Institutions of a Nilotic People.* Oxford: Oxford University Press; Fratkin, Elliot. 2004. *Ariaal Pastoralists of Kenya.* (2nd ed.). Boston, MA: Allyn and Bacon; Gray, Sandra. 2000. "A Memory of Loss: Ecological Politics, Local History, and the Evolution of Karimojong Violence." *Human Organization* 59.401–418, and Gray, Sandra, Mary Sundal, Brandi Wiebusch, Michael A. Little, Paul W. Leslie, and Ivy L. Pike. 2003. "Cattle Raiding, Cultural Survival, and Adaptability of East African Pastoralists." *Current Anthropology* 44(suppl):S3–S30. Also see McCabe, J. Terrence. 2004. *Cattle Bring Us to Our Enemies: Turkana Ecology, Politics, and Raiding in a Disequilibrium System.* Ann Arbor: University of Michigan Press, for an extreme conflict situation; and for a conflict-free one, see Moritz, Mark, Elizabeth Gardiner, Mark Hubbe, and Amber Johnson. 2019. "Comparative Study of Pastoral Property Regimes in Africa Offers No Support for Economic Defensibility Model." *Current Anthropology* 60:609–636.

36 Li, Wenjun, and Lynn Huntsinger. 2011. "China's Grassland Contract Policy and Its Impacts on Herder Ability to Benefit in Inner Mongolia: Tragic Feedbacks." *Ecology and Society* 16(2):article 1; Williams, Dee Mack. 1996a. "The Barbed Walls of China: A Contemporary Grassland Drama." *Journal of Asian Studies* 55:665–691; Dee Mack.

1996b. "Grassland Enclosure: Catalyst of Land Degradation in Inner Mongolia." *Human Organization* 55:307–313; Dee Mack. 2000. "Representations of Nature on the Mongolian Steppe: An Investigation of Scientific Knowledge Construction." *American Anthropologist* 102:503–519.

37 See, for instance, Kenin-Lopsan, Mongush B. 1997. *Shamanic Songs and Myths of Tuva.* Ed./tr. Mihály Hoppál. Budapest: Akadémiai Kiadó.

38 Stépanoff, Charles, Charlotte Marchina, Camille Fossier, and Nicolas Bureau. 2017. "Animal Autonomy and Intermittent Coexistences: North Asian Modes of Herding." *Current Anthropology* 58:57–81; On transhumance, see also Molnar, Zsolt. 2017. "'I See the Grass Through the Mouths of My Animals': Folk Indicators of Pasture Plants Used by Traditional Steppe Herders." *Journal of Ethnobiology* 37:522–541; Netting, Robert M. 1991. *Balancing on an Alp: Ecological Change and Continuity in a Swiss Mountain Community.* Cambridge: Cambridge University Press; and a classic work that includes reindeer herding, Vainstein, Sevyan. 1980. *Nomads of South Siberia: The Pastoral Economies of Tuva.* Edited and with introduction by Caroline Humphrey. Cambridge: Cambridge University Press.

39 Hastings, James Rodney. 1965. *The Changing Mile: An Ecological Study of Vegetation Change with Time in the Lower Mile of an Arid and Semiarid Region.* Tucson: University of Arizona Press; Turner, Raymond M., Robert H. Webb, Janice E. Bowers, and James Rodney Hastings. 2003. *The Changing Mile Revisited.* Tucson: University of Arizona Press.
For much more of the story of ranching in south Arizona and how it was saved and made sustainable in some areas (while wiping out in others), see: Sayre, Nathan F. 2002. *Ranching, Endangered Species, and Urbanization in the Southwest: Species of Capital.* Tucson: University of Arizona Press; Sayre, Nathan. 2001. *The New Ranch Handbook: A Guide to Restoring Western Rangelands.* Santa Fe, NM: Quivira Coalition; Sheridan, Thomas E. 1988. *Where the Dove Calls: The Political Ecology of a Peasant Corporate Community.* Tucson: University of Arizona Press; Sheridan, Thomas E. 2007. "Embattled Ranchers, Endangered Species, and Urban Sprawl: The Political Ecology of the New American West." *Annual Review of Anthropology* 36:121–154.

40 Sayre, Nathan F. 2002. *Ranching, Endangered Species, and Urbanization in the Southwest: Species of Capital.* Tucson: University of Arizona Press.

41 Sutton, Mark Q. 1988. "Insects as Food: Aboriginal Entomophagy in the Great Basin." *Ballena Press Anthropological Papers* No. 33. Socorro, New Mexico; Doi, Hideyuki, Remigiusz Gałęcki, and Randy Nathaniel Mulia. 2021. "The Merits of Entomophagy in the Post COVID-19 World." *Trends in Food Science & Technology* 110:849–854; Raheem, Dele, António Raposo, Oluwatoyin Bolanle Oluwole, Maaike Nieuwland, Ariana Saraiva, and Conrado Carrascosa. 2019. "Entomophagy: Nutritional, Ecological, Safety and Legislation Aspects." *Food Research International* 126:108672.

42 Cresson, Ezra T., Jr. 1934. "Descriptions of New Genera and Species of the Dipterous Family Ephydridae. XI." *Transactions of the American Entomological Society* 60:199–222 (see page 216, note 10).

43 Tomé, Daniel, and Cécile Bos. 2007. "Lysine Requirement through the Human Life Cycle." *The Journal of Nutrition* 137(6):1642S–1645S.

44 Maya agriculture here is summarized largely from Anderson, E. N. 2005. *The Political Ecology of a Yucatec Maya Community.* Tucson: University of Arizona Press; and Sutton, Mark Q., and E. N. Anderson. 2014. *Introduction to Cultural Ecology* (3rd ed.). Lanham, MD: AltaMira (Rowman and Littlefield). Also see Fedick, Scott (ed.). 1996. *The Managed Mosaic: Ancient Maya Agriculture and Resource Use.* Salt Lake City: University of Utah Press; Lentz, David, Nicholas P. Dunning, and Vernon L. Scarborough (eds.). 2015. *Tikal: Paleoecology of an Ancient Maya City.* New York: Cambridge University Press. Many other good works on this subject are available.

45 For example, see workingforestinitiative.com.

46 Summarized from much more detailed accounts in Anderson, E. N. 1988. *The Food of China*. New Haven, CT: Yale University Press, and Sutton, Mark Q., and E. N. Anderson. 2014. *Introduction to Cultural Ecology* (3rd ed.). Lanham, MD: AltaMira (Rowman and Littlefield), pp. 304–319.

47 Roosevelt, Anna C. 1999. "The Development of Prehistoric Complex Societies: Amazonia, A Tropical Forest." *Archeological Papers of the American Anthropological Association* 9(1):13–33; Woods, William I., Wenceslau G. Teixeira, Johannes Lehmann, Christoph Steiner, Antoinette WinklerPrins, and Lilian Rebellato (eds.). 2009. *Amazonian Dark Earths: Wim Sombroek's Vision*. Berlin: Springer; Clasby, Ryan, and Jason Nesbitt. 2021. *The Archaeology of the Upper Amazon: Complexity and Interaction in the Andean Tropical Forest*. Gainesville: University Press of Florida.

8

THE FORESTS THAT SUSTAIN US

"Under our current system, a forest has no value until it's cut down, which explains a lot about the root of our problems" (author unknown)

Among the fastest-diminishing resources on the planet are forests. The destruction of the world's forests is accelerating. Only about a quarter of the planet's forests are in good shape in 2023,[1] and protection is poor in most nations. By about 2050, there will be a shortage of lumber, and well before 2100, tropical forests will likely be gone. Increased demand will lead to rapid reversal of the gains of past decades. The three trillion trees in the world are probably less than half of what existed 5,000 years ago.[2]

Forests, both growing and old-growth, store between 25 and 33 percent of global carbon emissions. But forests are seriously threatened by deforestation and climate change. Deforestation is a prime cause of climate change, directly responsible for about 18 percent of it. Burning and decay of trees releases the stored CO_2, and the exposure of soil and lower vegetation releases more. The bare ground dries fast and reflects heat upward. The trees that once transpired vast amounts of water are no longer there, so fog, clouds, and rain diminish drastically. Rain diminishes downwind, and areas downwind of the massive deforestation of Amazonia are drying out.

An Assortment of Forests

In the high latitudes across the planet lie the boreal forests; they are largely evergreen, with vast expanses of pine, larch, and spruce. Fir is common in many areas, especially mountains. Massive forests survive in Scandinavia and Siberia. In the

DOI: 10.4324/9781003560326-8

western US and neighboring Canada, Douglas fir is the most common tree. A southward extension of boreal-type evergreen forests into the northwestern US produces the biggest trees in the world: the redwoods above all, but also enormous Douglas firs and pines. The world's tallest tree is a redwood in California that soars to 380.3 feet (116 m). In the far north, trees can be quite small, less than the height of a person. In Siberia, pines and larches dominate. Boreal forests extend to the highlands of Tibet, where tall trees also occur.

Moving to lower latitudes, the evergreen forests grade into broadleaved temperate-zone forests, dominated by maples, oaks, poplars (including aspens), and several other genera. Many formerly dominant species in these forests have been eliminated by pests, including chestnuts and ash trees in the eastern US and in China. Extremely diverse, species-rich temperate forests survive in Appalachia and western China but are rapidly dwindling. Southern hemisphere equivalents are the southern-beech forests of Patagonia and the eucalyptus forests of Australia. The latter produces some enormous trees. The Centurion tree in Tasmania (one of the provinces of Australia), at 327 feet (100 m), is the tallest tree in the world after the conifers. These forests grade into subtropical mixed forests; conifers often regain importance in these forests, particularly in North America and China.

In the deserts of western North America, forests of pinyon pine and juniper trees were common in the higher elevations, but much of those forests have been destroyed by wood cutters, miners, and by ranchers and government agencies making pastures. Equivalent areas in Asia have junipers and pines. In central Asia there is a tree, the saksaul (*Haloxylon ammodendron*), that grows in areas with only a couple of inches of rain a year.

In the equatorial regions, tropical forests are divided into two broad types. The tropical rainforest is world-famous for its beauty, variety, biodiversity, and value for blotting up carbon and producing oxygen. Far less recognized are the dry and seasonal forests. They take up about as much land but are less photogenic. Trees are shorter. Animals are more camouflaged. The long, dry seasons turn the forests gray-brown and nearly leafless, and they bake in the sun. A classic description of the Yucatan (in southeastern Mexico) dry forest is that "every plant has thorns and every animal bites or stings"—an exaggeration, but an understandable one, given the reality that confronts a traveler. (One whole group of acacias not only has huge thorns, but the thorns are hollow and house stinging ants.) Yet, these forests are almost as valuable to the planetary atmosphere as the rainforests. They have about as much biodiversity. And they have their own quiet beauty, one that requires the viewer an appreciation of subtle shades. One learns to appreciate shades of gray, rust, tan, and earth-tones.

Another type of tropical forest is the cloud forest, a realm of short, densely packed trees growing at high altitudes, where clouds keep the landscape usually moist. Cloud forests appear around 1,500–3,000 meters in elevation, depending on the mountains in question. These forests are almost impenetrable to humans and thus tend to be little known scientifically—realms of mystery and strangeness.

Scientists fight their way through nightmarish conditions, discovering whole ranks of new birds, insects, and plants.

All these forest types are divided into countless local biotic communities. Every tree has its own requirements, as do the smaller plants and the animals. They exist in a labyrinth of complex interactions, responding to competition and cooperation from each other. Fungal webs in the root systems usually weave trees together. These allow trees to share nutrients; carbon fixed by one species can be incorporated by trees of other species. They also transmit chemical messages from tree to tree, allowing them to communicate about as well as the simplest animals. The message—through fungi and aerially released stress compounds—is usually "Somebody is eating me! Watch out!" Other trees respond by mobilizing anti-herbivore compounds.

The fungal community in the roots also breaks down dead matter in the soil, liberating nitrogen and minerals for growth. This allows trees to grow faster, thus taking up more carbon. These fungal nets have enormous importance in driving and maintaining species diversity, but they also produce products beloved of gourmets, notably mushrooms, including the matsutake that has become so important in world trade.

Animals support forests and are vital to forest communities. Herbivores and seedeaters thin out plants. This is one reason for the incredible diversity of trees in the tropical rainforests, where a hectare (2.47 acres) can host 200 species of trees; herbivores quickly discover any group of trees of the same species and eat enough to keep the group small. On the other hand, extreme conditions can create one-species stands even in tropical rainforests. In Malaysia, the gelam tree is the only species that would grow in extremely acidic, mineralized soils. Few herbivores eat this tree, and no other trees can grow in those conditions, so it monopolizes the landscape.

Deforestation

Trees provide innumerable benefits. They store carbon, produce oxygen, help to mitigate the increasing heat, are homes to many other species, and anchor soils from erosion. Humans use trees for firewood (still widely used), building material, pulpwood for making paper, for shade and beautification, plus the sap is used for a variety of purposes (e.g., rubber and pancake syrup), and nuts for food. It is simply a bad idea to destroy such a resource.

The Wood Old Days

People have cut down trees since there were people and trees. Early on, the technology for tree cutting consisted of sharp stones or burning through the trunk. Either method was laborious and time-consuming, meaning that relatively few trees could be cut down—too few to really damage a forest. A valuable history of deforestation is *Deforesting the Earth: From Prehistory to Global Crisis* by Michael Williams.[3]

From it, we learn that in Europe, farmers arrived about 8,000 years ago and encountered a region almost completely covered by forest. They took to clearing the forest with chipped stone axes to make fields and pastures, while more trees were felled for building materials and firewood. The introduction of polished stone axes, followed later by copper and bronze axes, greatly increased the efficiency of tree cutting and accelerated the process of deforestation. The impact of the farmers' colonization of Europe also included extensive landscape modifications, alteration of hydrological systems, degradation of native species, and the destruction of the indigenous hunter-gatherer societies. By 3,000 years ago, farmers had deforested and transformed much of Europe. The remaining forests today are now often protected, but some old-growth forests, especially in eastern Europe, are still under major threat.

In some cases, deforestation and drought proved catastrophic. In addition to the deforestation, the first farmers extensively altered the landscape, modified the streams and rivers, and substantially degraded native ecosystems. Finally, the farmers destroyed the indigenous hunter-gatherer societies. Plant diseases, notably Dutch elm disease, also affected the Neolithic forests.

At the end of the neolithic in northwestern South Asia (today's Pakistan), the large sedentary farming villages in the Indus River Valley were dependent on the annual floods of the Indus River. They then began to coalesce into large cities, the two best-known being Harappa and Mohenjo-Daro, the beginnings of the Indus Valley state.

However, at about the time the Indus Valley state formed, rainfall appears to have lessened (drought), suggesting the possibility that the decreasing water supplies resulted in the coalescing of communities into the newly built cities. Subsequently, substantial deforestation along the headwaters of the river led to flooding that overwhelmed the farming system. Eventually, the continuing drought coupled with deforestation seems to have resulted in the abandonment of the region and the demise of the Indus Valley state.[4]

The conflict between conservation and exploitation is as old as record-keeping. Gilgamesh and his follower Enkidu fought a demon-like keeper of the cedar forest to get at the timber.[5] A more realistic picture is found in the Hebrew Bible. The king of Israel orders "a letter unto Asaph the keeper of the king's forest, that he may give me timber to make beams for the gates of the palace…and for the wall of the city, and for the house that I shall enter into" (Nehemiah 2:8). Kings protected the forests, keeping them under heavy guard. Similar records stretch from China to the pre-Columbian Americas.

Rapa Nui (aka Easter Island) in the southeastern Pacific Ocean is an interesting case. Polynesian explorers discovered and settled Rapa Nui some 750 years ago. They established a prosperous farming and fishing economy, founded settlements, and grew their population. Soon, people began to quarry and carve large stone statues called moai, almost a thousand of which are known. These statues weighed as much as 270 tons and had to be moved from the quarries to their destinations. It is assumed that the transport and erection of the moai would have required rope and logs. Pollen analysis indicates that at the time people first arrived at Rapa Nui, there was an extensive forest of trees, shrubs, and other plants from which the rope and logs could be harvested. Some researchers believe that the inhabitants of Rapa Nui overexploited their forest to the point that they could no longer support their society, with a resultant loss of population and abandonment of settlements. Adding to the damage were rats (introduced by Polynesians) that ate the seeds of the trees. The appearance of Europeans in 1722 made matters worse with disease and slave-trading decimating the population. Bad management all over.[6]

The Not-So-Wood New Days

There are about five billion acres of forest today, much of which is severely degraded or deforested and barely regrowing. We are seemingly chomping at the bit to finish the job. About two billion acres could be restored to full value.[7] About 223 million acres of forest will be lost by 2050 due to deforestation and climate change (e.g., dying due to drought or destroyed in wildfires).[8] Some one-third of tree species are threatened with extinction, and some species are down to a few individuals. Others are trapped in small areas that will eventually be destroyed by the warming climate.

Logging continues to be a contributor to the loss of forests. The temperate-zone forests were largely gone by 1900. Some recovery has occurred since but has not replaced much of the timber and nontimber products supply. Europe, actively reforesting since the 19th century, has reversed course and is logging heavily, although there is some movement back to protection. The US also reforested some land east of the Mississippi in the mid- to late-20th century, after a peak of deforestation in the 1920s but has since cut heavily into the forests in Western states in the late 20th century. Clear-cutting (Figure 8.1) is a major issue.[9] The clear-cutting of these forests is poorly known by the public since the loggers leave a swath of trees next to highways so the public does not see the devastation. Some of the forests that recovered after the 1920s in the east have been turned into pine plantations. Boreal forests are in better shape, but massive logging and rampant fires have destroyed many millions of acres of these forests in the 21st century. Cutting down millions of trees each year for Christmas and then throwing them away doesn't help.

The pulp and paper industry still consumes whole forests. Millions of trees are cut annually to make toilet paper. We know how to make paper out of fast-growing bamboo, hemp, and grasses; only entrenched interests—subsidized, as always—keep us deforesting.

Into all of this mess comes climate change. Drought weakens or kills trees, making them susceptible to insects and disease. The insects, in turn, no longer freeze in winter and thrive in warmer climates and less well-watered trees.

FIGURE 8.1   Clearcutting in Saint-Victor-Montvianeix, Auvergne, France. © Marie-Lan
              Nguyen, CC-BY-2.5, Wikimedia Commons).

Bark beetles—which farm fungus in burrows in the trees—have killed millions of
acres of forest in the US and Canada. Dry forest burns easily, causing pollution,
property damage, environmental damage, and release of carbon stored in the trees.
Wildfires are getting much more serious and deadly, even in Hawai'i, where the
Lahaina burn of 2023 was the deadliest fire in modern US history. Warmer and
drier weather combined with introduced weedy grasses that were highly flammable add to the problem. The fire became a catastrophe because of human neglect:
escape routes, water supplies, hospital protection, and indeed every safety measure
had been neglected despite the warning given by a smaller such fire four years
earlier. The old cry of "too much taxpayers' money" led to cutting back on safety,
which led, tragically, to fewer taxpayers.

Logging is more extensive than ever in the tropics. Huge subsidies for the conversion of forests to agriculture still exist in some tropical countries. As in the fossil
fuel and fishing industries, subsidies drive the real and extreme evils everywhere.
Governments are paying for the destruction of the planet. Since the loggers and
ranchers depend heavily on these subsidies, they invest a great deal of that money
in lobbying and outright bribery to increase the subsidies and to increase allowable
deforestation.

Clearing forests for cattle ranching, coffee plantations, local food production,
and urbanization led to the virtually total destruction of the Atlantic forests of
Brazil. Whole species of animals survive only in the trees left for shade-grown
coffee plantations. Logging, roadbuilding, mining, small farms, and other

problems have destroyed most of the tropical forests, especially in Indonesia and Brazil. Massive government support for ranchers, loggers, miners, and oil drillers, combined with indifference or active hostility to smallholders and Indigenous people, have been the norm for most of modern history. This reached extreme levels in the US in the late 19th century, in Brazil in the 21st, and in other places at various times. Canada and several Latin American nations have recently changed their tunes.[10]

The conversion of forests to farms and pastures is a major cause of deforestation. Pastures for cattle ranching account for about half of the deforestation in the tropics. The demand for beef by US markets is a major cause of deforestation in Brazil. Forests are also cleared for farms. However, many of the fields made from tropical forests have poor soils and are abandoned in a few years, with new fields cut to replace them. Speculative land clearing, land tenure issues, and farming-related fires spreading to adjacent forests are also issues. Reforestation under such circumstances is difficult.

Around 15 million acres of forest are cleared annually to make fields for oil palms, soybeans, rubber, cocoa, coffee, rice, corn, and cassava. Many of the world's tropical forests are either converted to oil palm or under some threat. Formerly confined to west Africa, the palm spread early to Indonesia, where it has led to total destruction of natural forest over many millions of acres. Cultivation is now spreading to Latin America and to other parts of Africa. The oil palm is at least a tree and fixes some carbon, but oil palm plantations have virtually no biodiversity and eliminate larger wildlife and other benefits. Islands of native forest left in oil palm plantations help but are not a solution.

Coffee has been a similar agent of destruction in higher woodlands, but shade-grown coffee is far less damaging to the ecology than sun-grown or most other species. In addition, coffee does not fix carbon as well as many of the trees it replaces.

A final blow is the illegal narcotics trade. Worldwide, but especially in Central and South America, drug gangs have been moving into forests to hide their operations, grow marijuana and opium, and increasingly to exploit illegal logging, mining, and other destructive activities.

Climate change is threatening all forests, but especially the boreal ones since the high latitudes are warming faster than elsewhere. Forests dry and burn, and vast areas of Siberia, Canada, and cold-weather forests in mountains of the US have been lost to fires in the last ten years. The warming allows pine beetles, emerald ash borers, and other boring insects that carry fungal infections to flourish and attack the trees. Then the trees die; they burn more easily, in turn liberating yet more carbon, and the vicious cycle goes on.[11]

The forest zones are shifting to higher and higher latitudes as the planet warms. However, many species are hopelessly unable to shift their ranges fast enough to avoid extinction. Some are stranded on mountaintops and have nowhere higher to go. Others are limited by soil fertility and could be increased by fertilizing, especially with phosphorus, which is often deficient in soils, but others require poor soils and are killed by artificial fertilization.

Finally, world trade has internationalized tree pests. The US reveals a particularly serious case since it imports specialty timber and wood products from all over the world. The chestnut blight arrived from Asia in the early 20th century, killed over a billion chestnut trees, and continues to destroy what was the most important and valuable tree on the continent for nuts and timber. More recently, a whole host of devastating pests have come in. The emerald ash borer, a small beetle that carries fungal diseases, has reduced six species of ash from major dominance and abundance in the eastern woodlands to rare or even endangered status. The woolly adelgid kills native conifers in the east; the white pine blister rust eliminates white pines in the west, thus endangering the animals dependent on their seeds, from mice to grizzly bears.[12]

Expanding human populations and economies are running up against rapidly diminishing amounts of wood, especially quality timber. Prize types of wood are already in short supply, commanding high prices. Madagascar rosewood, for example, is in enough demand to overwhelm the small resources of that country's protective services; illegal cutting is out of control.[13] The illegal logging of teak, especially in India and Myanmar, is another example. Even in Canada, with vast timber supplies, good knowledge of forestry, and good enforcement capacity, rampant logging is wiping out forests everywhere, especially the high-quality old growth of the west.

Regrowing Forests

The most direct and simple form of increasing trees and blotting up more atmospheric carbon is simply letting the forests regrow. Forests at high elevations in cold mountains grow almost imperceptibly slowly. Desert-edge woodlands (which are very extensive worldwide) also usually grow slowly, but some can grow surprisingly fast if they get reasonable water. Forests in the tropics may shoot up like weeds; some tropical trees, such as the fire-following Cecropia of tropical America, grow 4 meters or more in a year. Lowland Southeast Asian forests are notably fast-growing, and the less degraded they are, the better they do in regeneration.[14]

Forests have regrown before. The most spectacular success was that of Japan under the Tokugawa regime. The country went from largely deforested after the long and terrible civil wars of the 16th century to 90 percent forested in the 19th century. It is still largely forested. The story is told in Conrad Totman's book *The Green Archipelago*.[15] The Tokugawas did it by a simple and extremely effective, but merciless, approach: they ordered every forest community in Japan to grow trees, at least enough for government use. Headmen and loggers who failed were in serious danger of beheading. This was typical of Tokugawa approaches to conservation in general: best of intentions and the worst (at least for offenders) of enforcement methods. Fish, and for a while even dogs, were protected by this no-nonsense approach. Japan is now an ecologically well-off and sensitive nation, no longer using the death penalty.

The US hit a low point of forest area in the 1920s, since which forests have rather erratically increased because of the abandonment of marginal farmland, especially in New England, Appalachia, and much of the South. Often, trees are more valuable than the crops they replaced. For example, cotton has given way to pine plantations in much of the Old South, simply because the pines pay better.

An earlier, greater, and far more tragic regrowth resulted from the catastrophic decline in the Native American population after 1492. Many forests that had been maintained by Native people simply reverted to wild conditions. Farmlands became forested again. In the Caribbean, major regrowth took place on the islands and other areas that lost their large agricultural populations.

Europe has also seen massive regrowth since the 19th century. Unfortunately, these increases are more than offset by decline in the tropics and in the boreal forests, where logging, mining, and fire have been devastating. The same had happened earlier in Denmark, where trees for timber, windbreaks, and erosion control replaced many farms from the early modern period onward.[16] Much of Europe has rewilded to the point that wolves have returned. A newly proposed (2023) law in Europe, if passed, would require some 20 percent of farmland to revert back into forest.

Forest regrowth would enable considerably more carbon to be captured and stored. The most dramatic potential is in areas, Malaysia, Indonesia, and the Philippines, that have been extremely deforested. Worldwide, tropical dry forests are a huge source of possibilities. So are the northward-expanding boreal forests, as climate change pushes the tree line north and allows faster growth for forests in general. Parts of Europe and the south-central US could also grow a great deal more.

Given the problem with agricultural soil degradation noted above, the tendency is to reforest with single species of fast-growing trees that thrive on poor soil. Eucalyptus and pine are the worldwide favorites. But these do not bring back the biodiversity or local benefits of native forests. Worse, they are highly fire-prone, and monocultures have become deadly fire sources in the US, Portugal, Madagascar, South Africa, and many other countries. While this is generally a bad idea, it may be locally necessary in degraded soil situations.

Reaching Sustainable Use

Small communities and local polities have been maintaining forests for thousands of years. Management by local users is generally good since the communities depend on the wood, leaves, fruit, understory herbs, and other valued products. The greater the involvement by local users, the better the management. Studies show that communities with high levels of local management succeed in maintaining healthy forests. To illustrate, the forests managed by the federal government (the US Forest Service) are generally adversely impacted by fire, overlogging, heavy grazing, excessive road and dam building, excessively heavy recreational use, and other problems, while private and community woodlots flourish if carefully tended.

By far the best short-term hope for dealing with the world GHG crisis is through planting trees for use. In the long run, we must convert away from fossil fuel use, but right now we must blot up carbon, and forests do that. Bamboos and mangroves fix enormous amounts of carbon, and economic forests can be developed by planting fruit trees and valuable wood species.[17] The Chinese and Southeast Asians are doing this, but too few others are. A crash program in world tree planting, similar in scope to the mobilization for World War II, is needed.

Mexico's traditional communities maintained enormous areas of forest for many reasons, ranging from timber and firewood to religious veneration. These forests are widespread and constitute a major resource. Unfortunately, they have been rapidly disappearing as communities privatize land, expand agriculture and logging, succumb to global climate change, and otherwise suffer dubious fates. Some successes have occurred where people maintain some control over the lands and can work with foresters and environmental workers, but the task is daunting. In the Yucatan, the Maya there protect forests very well. They cut firebreaks, protect valuable trees from burning, manage regrowth to create the forests they want, and caretake existing forests to maintain useful species and biodiversity.[18] Other successes are found in the magnificent pine forests just north of Oaxaca city. If you can get to Google Maps, look at central Quintana Roo, Belize, and southern Campeche for Maya forests, and at the green zone between Oaxaca city and Santa Catarina Ixtepeji for Oaxaca. Then compare the lunar landscapes in some areas not far off.

Using trees more intensively is also a way to save forests. In California, thinned and cleared trees are often simply burned or left. They could be used. Even in logged forests, most of a given tree is left as slash; this could be salvaged for chipped-wood products and the like, as Europe and China often do.

Developing forests of varied fruit trees is also a wonderful idea. It has been done for millennia by traditional people in species-rich forest areas. The Native American forest communities apparently all managed at least selective cutting to spare fruit trees from clearing. Many, probably most, groups planted and transplanted trees widely. This was true even in far northwestern North America, where agriculture was unknown and planting was confined to these trees and, locally, to tobacco and some root crops. Peoples of eastern North America planted pawpaws and selectively burned the forest to maximize acorn, hazelnut, chestnut, walnut, hickory, and other resources. Hickory trees were so productive that it was more efficient to walk over 4 kilometers to a hickory tree than to produce the same amount of food by farming, even accounting for the enormous weight of nuts one had to haul back to town. No wonder farming came late in eastern North America and many other forested areas.[19]

The great home of forest management, especially for food trees, was the tropics. Here, the Maya (ancient and contemporary) selectively left fruit trees when clearing forest, thus gradually enriching the forest over cycles of swidden cultivation. The entire Amazon rainforest has been profoundly shaped by similar practices as well as by outright planting. Nut and fruit trees are much more common than they would be without people. Not surprisingly, giving Indigenous people title to their land is an extremely effective way to save forests in Amazonia, since the Indigenous people use the forests while outsiders merely cut them down. Unfortunately, the Brazilian government under Bolsonaro did everything possible to throw Indigenous forests open to loggers and ranchers. By 2020, the whole east and south of the Amazonian forest belt were treeless except for tiny woodlots and reserves too small to preserve much biodiversity.[20]

Sacred Groves

Worldwide, one of the main devices for saving trees has been religion. Ancient Europe had sacred forests or woods, from Greece and Rome to the Celtic and Germanic oak woods. Specifically sacred trees locally included birch (north Europe), hawthorn (Celtic), oaks (generally), laurel (Mediterranean), and others. Sacred groves are so universal that even the modern, secular Western societies preserve trees in churchyards and cemeteries. These often maintain shreds of native vegetation otherwise lost.

Public parks, national parks, and estate grounds are, to some degree, secularized descendants of sacred groves. They also recall the hunting parks of the nobility, and thus national parks and reserves are sometimes condemned by anti-environmentalists as "privilege" and remnants of class discrimination. Yet even earlier, the groves and woods that evolved into elite property were communal forests, often sacred ones. They were seized by the nobility in medieval times, part of the widespread encroachment on the commons that took place at the time.

Elsewhere, whole forests can be maintained for religious reasons. The Pure Crystal Mountain of Tibet is preserved in its entirety for pilgrimage, veneration, and retreat. Sacred forests occur in Israel, Africa, and Siberia. Ethiopia preserves groves in church land; in heavily farmed parts of the country, these may be the only bits of native vegetation left. The same is true in southern Madagascar.[21]

The idiom "tree-hugging," now often used in an irreverent and dismissive way to refer to environmentalists, has a deadly serious meaning: desperate Indian farmers literally hugged their sacred trees to save them from ruthless, dubiously legal logging operations. The farmers believed that the loss of the trees would mean certain doom for the region. A similar group in Rajasthan tries to preserve its limited forests, which are desperately needed to bring rain and retain water, thus relieving drought in that desert province. A major book, *People Trees* by David Haberman, describes the whole movement to save groves through religion in India.[22]

In southern Morocco, groves of juniper, oak, and argan are often conserved as sacred, sometimes managed by descendants of the saint or holy person honored by the grove.[23] The argan tree is valued for its extremely high-quality and tasty oil from its seeds. This oil is now used internationally as a cosmetic, but in its home, it is valued as food. Among other things, it is mixed with honey and crushed almonds to produce what is locally sold as "Moroccan Viagra" (but there is no evidence that it works for that purpose). It grows in desert environments, where it must be carefully nurtured and tended to protect it. It is the tree the goats are climbing in all those cute photos of goats in trees, but the goats must be carefully managed to avoid browsing the trees to death. Caretaking these trees is both a religious duty and a government project.

In Thailand, reverence for sacred trees and groves is evident to even the most casual visitor. Even street trees are often worshiped. Protection of sacred trees from loggers and urban construction is still a major force in Thai political life. It has brought Buddhist monks and modern environmentalists and ecologists together in efforts to save forests, often by designating them as sacred.[24]

Individual trees are worshiped in many areas. Throughout India, Southeast Asia, and southern China, these are apt to be banyans—tropical figs with multiple roots that descend from trunks and branches to the ground, making a forest from one tree. The Buddha was said to have been enlightened under a banyan. Other venerated trees of Asia include walnuts, sycamores, cypresses, ginkgos, and dozens of other species. Birch trees are widely considered sacred throughout northern Eurasia and into North America. Red cedar (actually a cypress) is particularly venerated on the Northwest Coast of North America, but in that area all trees were traditionally respected. Some people still refuse to take a piece of bark from a tree without asking its permission and then thanking it.[25]

Forests without protection from spiritual, religious, or cosmological justification often fail to survive. They are particularly unlikely to preserve rare or unusual trees and forest types. A major exception, noted above, of considerable interest is the game park as an institution. This is the original "paradise"—paradaizu, ancient Iranian for a walled garden or hunting park, an idea probably developed in ancient Mesopotamia. The idea spread throughout Eurasia. Vast hunting parks abounded in early times but shrank, slowly or rapidly, as farming and cities grew and encroached on them. The idea survives today in modern game parks and hunting clubs. These were already known in ancient times and have grown since.

Today, protection is far more widespread. The national forest concept, developed from tree management ideas in Germany and elsewhere, caught on in the US and then worldwide. Protection of some sort is found in all countries with significant forests. However, few have adequate protection. Small areas do not adequately conserve biodiversity. Reserves are easily compromised or abolished by anti-environmental governments. Some areas of the world have done well, but many of these areas are small and remote. Europe has done well, but its forests need more management; most have been heavily used for centuries. Some countries are still logging heavily, notably Sweden, and some have much old growth not well protected, notably Romania.

The Case of China

The core provinces of China, as well as Taiwan, were almost entirely forested 5,000 years ago. Only the northwest portion of the county was dominated by deserts and grasslands. Gansu province was largely desert, though even it had mountain forests. Extensive grasslands and brushlands existed in Shaanxi, Shansi, and neighboring bits of other provinces. High mountains reached above the tree line, or beyond the limits of adequate rain, in Sichuan and Yunnan. Otherwise, however, China was forested except for local prairies and marshes.

In the intervening millennia, China carefully protected many forests for religious and cosmological reasons, as sacred temples or feng shui groves. A timber industry developed in the Middle Ages, and China developed scientific forestry, anticipating the west by centuries. Most of the management techniques made famous by German foresters in the 19th century, such as even-age monoculture with selected stock, were developed much earlier in China. How much German foresters learned from China remains an open question.

Indigenous minority nationalities conserve forests in China, at least as well as those elsewhere. In particular, upland Southeast Asia has a unique combination of high population density, low levels of governance (until recently), and effective conservation of resources. This was combined with a region-wide ideology of conservation that involved respect for all lives, including the spirit beings of hills, forests, and indeed all places. The forests were held to be both necessary for real-world benefits and sacred to the spirits that dwelt in them. A balance had to be maintained between the forest spirits and the spirits of the human realm of villages and fields. Harming forests destroyed village safety both through obvious problems like fuel shortage and erosion and through the evil luck that offending the spirits would bring.

Conservation was inevitably imperfect, but much (if not most) of the damage was related to attempts to protect villages from outside forces, usually imperial states. One need only look at the satellite photographs on Google Maps to see the dramatic differences at the national borders of China, Laos, and Myanmar. The latter two countries have not been successful at dominating and expropriating small groups, and the forest is in good shape. On the other hand, China has for thousands of years moved south, encroaching on local Indigenous groups, reducing them to oppressed-minority status. The region is now rapidly converting to rubber, oil palm, and other plantation crops.

A famous Chinese poem from around 140 BCE described in exaggerated detail the royal hunting park but goes on to praise the emperor for throwing much of it open to cultivation by farmers desperate for land. The contrast of mental imagery between lush hunting and sober farming shows an early manifestation of the Chinese ambiguity between love of the wild and of natural scenes, but the endless pressure to turn them into farm and city.[26]

Thus, by 1950, China's magnificent and extensive forests were almost gone, surviving only in the far south and far northeast. Reforestation under the

People's Republic has dramatically changed that, restoring mountain forests almost everywhere, though not doing much for the lowlands.

Burn, Baby, Burn

Fire is the cleaner of the land, the renewer of life, and is a normal part of most natural communities, especially forest communities, except in extremely wet areas such as the Pacific Northwest and the wettest tropical rainforests. Most forests depend on fire for periodic renewal and opening of space. The ash carries accumulated nutrients from years or decades of growth and thus fertilizes new stands.

Many trees require fire to sprout their seeds. Whole species-complexes of pines have adapted to this lifestyle. They require fire to open their cones and liberate the seeds. One such is the lodgepole pine, which often occurs in pure stands all the same age, recalling a fire that allowed them to dominate. Some populations of this tree grow senile after about 50 years and need another fire to regrow the stand, though other populations of lodgepoles live for centuries. Other plant species simply need open space. Many have seeds that are stimulated to sprout by chemicals released in forest fires. Forests are adapted to particular fire intervals; some "want" a fire every 50 years, some every 100, some every 200. When humans either prevent forest fires or drastically decrease their frequency, the forests decline.

Natural fires periodically consume the fuel on the ground so that the fuel cannot build up to the point of allowing a catastrophic fire to occur. A recently burned area can be quite productive in its abundance of new growth. In contrast, Western philosophy is that fire is bad and should be extinguished as fast as possible. This suppressive approach allows the fuel within the forest (called duff) to build up such that when it does catch fire, the resulting catastrophic fire cannot be extinguished and will destroy the forest. Such fires have devastated California over the last century or so. Thus, the well-intentioned policies meant to preserve the forest (and the people living within it) are the cause of its destruction.

More recently, some forest managers have recognized this issue and now try to let fires burn to reestablish the natural cycle. This was the case in 1988 when huge fires engulfed Yellowstone National Park (especially lodgepole pine forests). The fires were allowed to burn naturally, and the forest has since recovered and probably won't burn again for decades. Much more common are small, controlled burns of the sort carried out by traditional people in virtually every forested area of the Earth. Such small-scale controlled burning is being brought back in Australia and parts of the US, among other places.

Throughout prehistory, people have used fire to clear away dry, dead understory and grasses, reduce plant competition, and eliminate thorns and dangerous animals. Such fires were usually timed and managed so that they do minimal damage to forests and other long-lived resources. Sometimes fires were set to increase the harvest of seed plants. In addition, the resultant new growth would attract some game animals (e.g., deer), increasing hunting success.

Burning also may be related to religion, such as world renewal activities. Among the Gagadju in north-central Australia (as seen in the documentary film Twilight of Dreamtime), the burning of the river plain signals the renewal of an annual cycle of life. Without the ceremonial intervention by "proper" people, as caretakers of the land, this cycle would be broken and life would cease to exist. In fact, the burning does indeed renew the land. Controlled burning by Indigenous Australians has been reintroduced at Uluru (Ayers Rock) to save wildlife that was dying out for lack of it.

In short, revival of age-old local methods of managing forests can successfully restore forests if implemented. Modern scientific forestry can add to this and bring our trees back. As in other cases, however, entrenched interests that profit from massive and destructive deforestation have too much political power. There is nothing inevitable about this. We could save forests despite climate change and other threats.

Notes

1 Li, Wang, et al. 2023. "Human Fingerprint on Structural Density of Forests Globally." *Nature Sustainability* 6:368–379.

2 Hirons, Andrew D., and Peter A. Thomas. 2018. *Applied Tree Biology*. Hoboken, NJ: Wiley Blackwell; Crowther, T. W., et al. 2015. "Mapping Tree Density at a Global Scale." *Nature* 525:201–205.

3 Williams, Michael. 2003. *Deforesting the Earth: From Prehistory to Global Crisis*. Chicago: University of Chicago Press. Superior broad history.

4 Dutt, Som, Anil K. Gupta, Bernd Wünnemann, and Dada Yan. 2018. "A Long Arid Interlude in the Indian Summer Monsoon during~ 4,350 to 3,450 cal. yr BP Contemporaneous to Displacement of the Indus Valley Civilization." *Quaternary International* 482:83–92.

5 Gardner, John, and John Maier. 1984. *Gilgamesh, Translated from the Sin-Leqi-Unninni Version*. New York: Alfred A. Knopf.

6 Learn more about Rapa Nui in *The Archaeology of Rapa Nui (Easter Island)* by Terry L. Hunt and Carl Lipo (2018) in The Oxford Handbook of Prehistoric Oceania, Ethan E. Cochrane and Terry L. Hunt, eds., pp. 416–449. Oxford University Press, Oxford, United Kingdom.

7 Bastin, Jean-Francois, Yelena Finegold, Claude Garcia, Danilo Mollicone, Marcelo Rzende, Devin Routh, Constantin M. Zohner, and Thomas W. Crowther. 2019. "The Global Tree Restoration Potential." *Science* 365:76–79.

8 Bastin, Jean-François, et al. 2017. "The Extent of Forest in Dryland Biomes." *Science* 356:635–638; Bauman, David, et al. 2022. "Tropical Tree Mortality Has Increased with Rising Atmospheric Water Stress." *Nature* 608:528–533.

9 Ceccherini, Guido, et al. 2020. "Abrupt Increase in Harvested Forest Area over Europe after 2015." *Nature* 583:72–77.

10 For tropical forests, see Bebbington, Anthony J., Denise Humphreys Bebbington, Laura Aileen Sauls, John Rogan, Sumali Agrawal, Cesar Gamboa, Aviva Imhof, Kimberly Johnson, Herman Rosa, Antoinette Royo, Tessa Toumbourou, and Ricardo Verdum. 2018, "Resource Extraction and Infrastructure Threaten Forest Cover and Community Rights." *Proceedings of the National Academy of Sciences* 115:13164–13173; Dean, Warren. 1995. *With Broadax and Firebrand: The Destruction of the Brazilian Atlantic Forest*. Berkeley: University of California Press; Neslen, Arthur. 2015. "Subsidies to

Industries that Cause Deforestation Worth 100 Times More than Aid to Prevent It." *The Guardian*, online, March 31, http://www.theguardian.com/environment/2015/mar/31/subsidies-to-industries-that-cause-deforestation-worth-100-times-more-than-aid-to-prevent-it?CMP=share_btn_fb; Painter, M., and William Durham (eds.). 1995. *The Social Causes of Environmental Destruction in Latin America*. Ann Arbor: University of Michigan Press; Peluso, Nancy Lee. 1992. *Rich Forests, Poor People: Resource Control and Resistance in Java*. Berkeley: University of California Press.

11 Kurz, Werner A., C. C. Dymond, Graham Stinson, G. J. Rampley, E. T. Neilson, A. L. Carroll, Tim Ebata, and Les Safranyik. 2008. "Mountain Pine Beetle and Forest Carbon Feedback to Climate Change." *Nature* 452:987–990.

12 Fei, Songlin, Randall S. Morin, Christopher M. Oswalt, and Andrew M. Liebhold. 2019. "Biomass Losses Resulting from Insect and Disease Invasions in US Forests." *Proceedings of the National Academy of Sciences* 116:17371–17376.

13 Barrett, Meredith A., Jason L. Brown, Megan K. Morikawa, Jean-Noel Labat, and Anne D. Yoder. 2010. "CITES Designation for Endangered Rosewood in Madagascar." *Science* 328:1109–1110; Irwin, Aisling. 2019. "Cops and Loggers." *Nature* 568:19–21.

14 Banin, Lindsay, et al. 2014. "Tropical Forest Wood Production: A Cross-Continent Comparison." *Journal of Ecology*, doi 1111/1365–2745.12263; Bastin, Jean-François, et al. 2017. "The Extent of Forest in Dryland Biomes." *Science* 356:635–638; Pugh, Thomas A. M., Mats Lindeskog, Benjamin Smith, Benjamin Poulter, Almut Arneth, Vanessa Havard, and Leonardo Calle. 2019. "Role of Forest Regrowth in Global Carbon Sink Dynamics." *Proceedings of the National Academy of Sciences* 116:4382–4387; Roebroek, Caspar T. J., et al. 2023. "Releasing Global Forests from Human Management: How Much More Carbon Could Be Stored?" *Science* 380:749–753.

15 Totman, Conrad. 1989. *The Green Archipelago: Forestry in Preindustrial Japan*. Berkeley: University of California Press. See also his book of 1995, *The Lumber Industry in Early Modern Japan*. Honolulu: University of Hawai'i Press.

16 Hecht, Susanna B., Kathleen D. Morrison, and Christine Padoch (eds.). 2014. *The Social Lives of Forests: Past, Present, and Future of Woodland Resurgence*. Chicago: University of Chicago Press.

17 Hawken, Paul. 2017. *Drawdown: The Most Comprehensive Plan Ever Proposed to Reverse Global Warming*. New York: Penguin.

18 Anderson, E. N. 2005. *Political Ecology of a Yucatec Maya Community*. Tucson: University of Arizona Press Bray, David Barton, Leticia Merino-Pérez, and Deborah Barry (eds.). 2005. *The Community Forests of Mexico: Managing for Sustainable Landscapes*. Austin: University of Texas Press; Mathews, Andrew. 2011. *Instituting Nature: Authority, Expertise and Power in Mexican Forests*. Cambridge: MIT Press.

19 The finding on walking to a hickory grove was made by Zeanah, David W. 2017. "Foraging Models, Niche Construction, and the Eastern Agricultural Complex." *American Antiquity* 82:3–24, who actually lugged all those hickory nuts to town. For the rest of the paragraph, see Armstrong, Chelsey Geralda, Jesse E. D. Miller, Alex C. McAlvay, Patrick Morgan Ritchie, and Dana Lepofsky. 2021. "Historical Indigenous Land-Use Explains Plant Functional Trait Diversity." *Ecology and Society* 26:6; Delcourt, Paul, and Hazel Delcourt. 2004. *Prehistoric Native Americans and Ecological Change: Human Ecosystems in Native North America Since the Pleistocene*. New York: Cambridge University Press; Doolittle, William. 2000. *Cultivated Landscapes of Native North America*. New York: Oxford University Press.

20 Again, use Google Maps' satellite views. See Albert, James S., et al. 2023. "Human Impacts Outpace Natural Processes in the Amazon." *Science* 379:348; Balée, William. 2013. *Cultural Forests of the Amazon*. Tuscaloosa: University of Alabama Press; Flores, Bernardo M., and Carolina Levis. 2021. "Human-Food Feedback in Tropical Forests." *Science* 372:1146–1147. Also see Baragwanath, Kathryn, Ella Bayi, and Nilesh Shinde. 2023. "Collective Property Rights Lead to Secondary Forest

Growth in the Brazilian Amazon." *Proceedings of the National Academy of Sciences* 120:e2221346120; Laurance, William F., Mark A. Cochrane, Scott Bergen, Philip M. Fearnside, Patricia Delamonica, Christopher Barber, Sammya D'Angelo, and Tito Fernandes. 2001. "The Future of the Brazilian Amazon." *Science* 291:438–439; Levis, C., F. R. C. Costa, F. Bongers, M. Peña-Claros, C. R. Clement, and ca. 70 others. 2017. "Persistent Effects of Pre-Columbian Plant Domestication on Amazonian Forest Composition." *Science* 355:925–931.

21 Abbott, Alison. 2019. "Saving Ethiopia's Church Forests." *Nature* 565:548–549; Bodin, Örjan, Maria Tengö, Anna Norman, Jakob Lundberg, and Thomas Elmqvist. 2006. "The Value of Small Size: Loss of Forest Patches and Ecological Thresholds in Southern Madagascar." *Ecological Applications* 16:440–451; Haberman, David L. 2013. *People Trees: Worship of Trees in Northern India.* New York: Oxford University Press; Huber, Toni. 1999. *The Cult of Pure Crystal Mountain.* New York: Oxford University Press; Slicher von Bath, B. H. 1963. *The Agrarian History of Western Europe, 500–1850.* New York: St. Martin's.

22 Haberman, David L. 2013. *People Trees: Worship of Trees in Northern India.* New York: Oxford University Press.

23 Genin, Didier; Romain Simenel. 2011. "Endogenous Berber Forest Management and the Shaping of Rural Forests in Southern Morocco: Implications for Shared Forest Management Options." *Human Ecology* 39:257–269. The rest of this paragraph is based on ENA's research in Morocco.

24 Sponsel, Leslie. 2012. *Spiritual Ecology: A Quiet Revolution.* Santa Barbara, CA: Praeger (Imprint of ABC-CLIO); Darlington, Susan. 2013. *The Ordination of a Tree: The Thai Buddhist Environmental Movement.* Albany: SUNY Press.

25 Personal research by ENA, but see Lubi, Ian H., Steve J. Miller, and Stephen Polasky. 2022. "When and Where to Protect Forests." *Nature* 609:89–93.

26 Sima Xiangqing (Sima Xiangru). 1987. "Rhapsody on the Imperial Park." In *Wen Xuan or Selections of Refined Literature, Vol. II*, Xiao Tong, ed., translated by David R. Knechtges and Tong Xiao, pp. 73–114. Princeton, NJ: Princeton University Press. Sima was a romantic hero in old China; he eloped with a high-born lady and got away with it. The idea of ambiguity in China about the environment was developed by Thornber, Karen Laura. 2012. *Ecoambiguity: Environmental Crises and East Asian Literatures.* Ann Arbor: University of Michigan Press.

9

THE SEAS WE SAIL

"How inappropriate to call this planet Earth when it is clearly Ocean."
(Arthur C. Clarke)

Saltwater covers some 71 percent of the planet, generally known as the ocean or the sea. For convenience, the ocean has been divided into five large oceans (Pacific, Atlantic, Indian, Arctic, and southern/Antarctic) and seven major but smaller, and largely bordered by land, seas (Mediterranean, Black, North, Red, Baltic, Arabian, and Caribbean), plus numerous smaller named seas. All of these bodies of water are connected and contain gulfs, bays, channels, straights, deltas, and the like. The ocean has, for many millennia, been used for trade, transportation, and warfare to produce food and recreation. These uses continue today.

Life is dependent on the oceans and likely originated there, billions of years ago. The freshwater we all use ultimately comes from the oceans, water that evaporates from the oceans and falls on the land as rain or snow. This water flows back into the oceans to become rain once again, making a perpetual cycle.

The oceans are among the "lungs of the planet" and absorb and store vast quantities of carbon, perhaps some 30 percent of the carbon that has been emitted in the last few hundred years. However, the capacity of the ocean to store carbon is near its limit, and it cannot store much more. Unfortunately, the presence of so much stored carbon in seawater has caused the oceans to become more acidic, impacting sea life.

The oceans also absorb heat. This extra heat provides the energy for more and more powerful storms, such as hurricanes. Warmer water also changes ocean currents, bringing storms to new places, such as the hurricane that impacted southern California in the summer of 2023. Ocean water is circulated by currents, and

DOI: 10.4324/9781003560326-9

surface water temperatures influence atmospheric circulation. Thus, weather, and so climate, is largely dictated by ocean conditions.

Despite its importance, we know remarkably little about the ocean. We know it has currents, tides, winds, and storms; all information invaluable to sailors. We know the ocean is as deep as 35,000 feet in some places, and we have some understanding of its geology, such as plate tectonics (aka continental drift) and volcano hot spots. We know it contains life in astonishingly varied forms, from mammals to fish to squid, to algae, and many more. Importantly, we know we are dependent on the ocean, even if we do not act like it.

Health Check

An understanding of the health and status of the ocean is dependent on our knowledge of it. The ocean is vast and difficult to study. While the surface of the ocean is important and easily visible, the underwater part is far more difficult to explore and study, and the deeper one goes, the more difficult it becomes. One must generally have ships, specialized diving equipment, or remotely operated submersibles, and considerable funding. Thus, even the many dedicated ocean scientists have been able to compile information on only a small percentage of its expanse, and the majority of the ocean remains unexplored (it is said that we know more about space than the ocean). Much of the ocean bottom has been mapped (mostly for military purposes), and this mapping led to the discovery of the mid-Atlantic ridge (and others) that confirmed plate tectonics.

Despite this lack of knowledge, we are keenly aware of some major issues. We know the water in the oceans is getting warmer. We know that the water in the oceans is becoming more acidic. We know that the oceans are badly contaminated. We know that sea level is rising. And we know that the resources within the ocean are overexploited and in danger of collapsing. Further, we know that each of these issues results in adverse impacts to the oceans, life, climate, and ultimately humans and their societies. Many of the specific impacts are not well understood, but the general trend is clear.

Oceanic Heatstroke

The planet is getting hotter due to the insulating effects of GHGs. The oceans are also becoming warmer, and this fact has a number of consequences. First, the oceans can no longer serve as heat sinks to blot up heat generated by GHGs. Second, as water warms, it expands, and that alone accounts for a sea level rise of several feet. Third, the warmer water is melting the Arctic and Antarctic ice sheets (the Arctic will soon be ice-free in summer), and this results in further sea level rise. This also results in the disruption of currents, and the Gulf Stream is in danger of shutting down, which would bring cold to northern Europe but heat to the Mediterranean and drought to central Asia.

Fourth, warmer water can be detrimental to some marine life. In July 2023, the waters around Florida reached 101°F, the highest temperature ever recorded in continental US waters. It represents a full-scale catastrophe for coral reefs.[1] The corals become bleached as they eject the algae they depend on for food and energy. This usually means death for the reefs. There is no longer any hope for corals around Florida. Most of the Caribbean reefs are already gone. Elsewhere in the world, Australia's famous Great Barrier Reef is suffering, and corals everywhere are threatened.

Finally, warmer water impacts a variety of other marine life. The warmer the water, the less oxygen it can hold and this will force some species to move to cooler waters or die. Fisheries are thus affected, meaning food supplies are stressed. None of this is good.

Oceanic Heartburn

The ocean has always had a level of absorbed carbon, but recently absorbing even more carbon has resulted in a lowering of the ocean pH, making it slightly more acidic. This change affects the ability of marine organisms, such as corals and mollusks, to build and maintain their calcium carbonate structures. As a result, coral reefs are threatened across the world due to acidification, not to mention the increasing sea temperatures. Shellfish suffer the same problems. Since shellfish are an important food for people, this could impact global food supplies.

Oceanic Contamination

People have always used water to dispose of waste, often in rivers that empty into the ocean or directly into the ocean. In ancient times, before so many people and massive industries, this was not much of a problem since it was so small scale. Still, even in ancient cities on rivers, such as Rome, Paris, and London, the rivers were essentially sewers. Population growth and the industrial revolution changed the scale.

Until quite recently, there was little to no control over what was dumped into the ocean. Industrial waste, agricultural chemicals, garbage, and sewage was liberally dumped into lakes and rivers that flowed into the ocean and often directly into the ocean itself. Polluted fluids from factories were piped directly into waterways or pumped out to sea. Sewage was dealt with in the same way. For decades, New York City and Los Angeles put their trash and garbage on barges and dumped them in the ocean. While the garbage problem was "out of sight, out of mind," medical waste, including syringes, washed up on New Jersey beaches, keeping the issue visible. The federal government forced a halt to the New York practice in 1982 (Los Angeles had stopped much earlier).

Pollutants in rivers eventually reach the ocean and can cause problems. The Mississippi River flows into the Gulf of Mexico and once formed a region with an amazingly diversity of plants and animals, a haven for farmers and shrimpers.

However, over time, the agricultural chemicals, industrial waste, and sewage overwhelmed the ability of the Gulf to cope, creating a biological dead zone that runs from the mouth of the river west along to coastline almost as far as the Mexican border (Figure 9.1).

A similar situation has occurred where the Nile River empties into the Mediterranean Sea. Since the Aswan High Dam was completed in the 1960s, natural fertile silt no longer covers the fields, forcing farmers to use chemical fertilizers. Over the last 60 years, the accumulation of these pollutants in the Nile Delta has created a biological dead zone in the Mediterranean, effectively destroying the once productive fisheries there.

Other pollution originates from human activity in the ocean itself. Among the worst such environmental disasters are the periodic oil spills either from oil production facilities (e.g., the Deepwater Horizon blowout in 2010) or oil tankers (e.g., the Exxon Valdez spill in 1989). These accidents devastate marine ecozones, kill millions of animals, and could have been avoided with better regulation, maintenance, training, and supervision.

A timebomb of sorts lies at the bottom of the ocean in the form of sunken ships. While ships have been sinking for thousands of years, the wrecks of more recent vessels can be quite hazardous. Ships made of iron and steel and powered by fossil fuels contain many pollutants (e.g., oils and fuels) that leak into the water. If the ships were warships, they likely contain weapons, explosives, and chemicals that are not good for marine life. In other cases, sunken ships can form habitat for a variety of marine life, and some ships, once cleaned of pollutants, are purposefully scuttled to form such habitats.

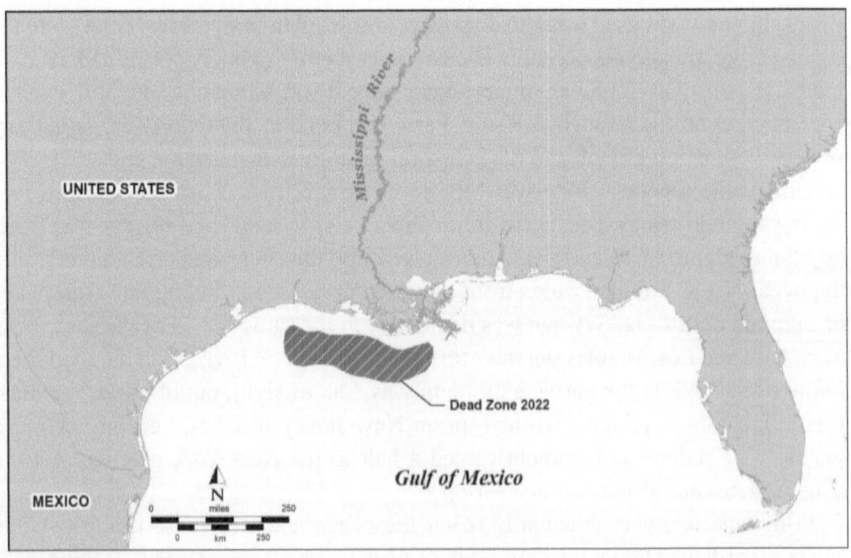

FIGURE 9.1 Map of the 2022 extent of the dead zone in the Gulf of Mexico.

Finally, much of the cargo now transported across the globe is shipped in standardized shipping containers loaded onto container ships and offloaded directly onto trucks or trains. The container ships transit the oceans and encounter a variety of sea conditions, some of which cause shipping containers to fall off the ship and into the ocean, usually to sink to the bottom. The contents of such lost containers might be hazardous, and some containers that do not sink become maritime hazards themselves.

A Plasticized World

Seemingly everything we see is made of plastic. And, everywhere we look, on land or sea, we find plastic trash. We currently produce some 350 million tons of plastics a year (that tonnage is growing), use it for some purpose, and then recycle or discard it. Plastics are made from oil and come in a bewildering range of types, from thin grocery bags to major industrial components. Plastics are now used for mostly everything.

Most plastics cannot be economically recycled into new plastic products. Those that cannot are discarded to end up either in landfills, illegally dumped, incinerated, or used as fuel (plastics, made from oil, burn well) in industrial processes such as cement manufacturing.

About three percent of plastic waste enters the oceans and can be found everywhere we look, even in the deepest places. Much of the plastic stays close to shore (Figure 9.2); some settles to the bottom of the ocean, while some disintegrates and is incorporated into sediments. A portion drifts out into the surface of the ocean to form gigantic "garbage patches" (Figure 9.3), such as the "Great Pacific Garbage Patch," an area the size of Texas located between the west coast of the US and Hawai'i. Efforts to clean up these garbage patches are underway by private organizations (governments do not seem to care).

Much of the discarded plastic waste breaks down into small fragments, called microplastics. These microscopic pieces of plastic are now almost everywhere. They are present in the guts of many of the animals we eat and are even in our own bloodstreams. For one depressing example, cigarette butt filters are made from microplastics, and five trillion (trillion with a T) are discarded into the environment each year.

In the ocean, microplastics are causing increasing damage to fish and other marine life. Reefs are being destroyed by both large plastic items, including fishing equipment, and microplastic bits.[2] Seabirds are dying in large numbers from plastics as well as from accidental capture in fishing operations; 90 percent of seabirds have some level of contamination, and 29 percent in one study had plastic in the gut.[3] Marine foodwebs are being destroyed. The so-called "biodegradable" plastics often just degrade into microplastics and wind up causing much more damage and pollution than if they had not "degraded."

How do we deal with this massive issue? Perhaps we can synthesize enzymes, especially from bacteria and wax moths, that can eat plastic.[4] Diana Kwon notes

FIGURE 9.2   Close-up of trash covering a tide line on a beach in Hawaiian Islands Humpback Whale National Marine Sanctuary. Photo by Matt McIntosh/NOAA.
Public Domain. (https://commons.wikimedia.org/w/index.php?search=trash
+on+a+beach&title=Special:MediaSearch&go=Go&type=image)

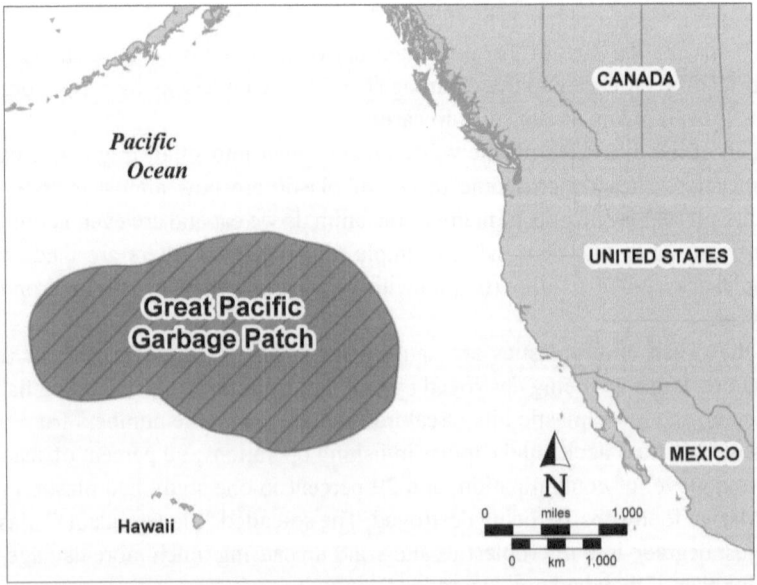

FIGURE 9.3   Map of the "Great Pacific Garbage Patch" off the western coast of North
America.

that remedies that work, to some extent, include charges for plastic items, often to be repaid if recycling occurs; higher taxes on plastic; bans on such items as plastic bags for merchandise; and other such incentives. More serious are plans to develop more recyclable forms of plastic, have more uniformity in plastic materials to make recycling easier, develop more organic methods for destruction (e.g., plastic eating bacteria), and to make rapidly biodegradable plastics that do not simply break down into microplastics.[5]

Not Just for Drinks

The ice found in glaciers contains a record of the environment stretching back some 800,000 years. Ice accumulates in layers, much like tree rings, and can be counted and dated. The ice layers contain various materials, such as ash, soot, dust, pollutants, and most importantly, bubbles of ancient air. Ice is sampled by taking cores, dating the various layers, and measuring what was in the ice at specific points in time. This record tells us the concentrations of GHGs in the air at any given time and is the library of the health of the planet.

Ice is important for other reasons. Ice and snow reflect solar radiation back into space, so the loss of ice makes things warmer. As the polar regions warm, the methane-soaked permafrost will melt, releasing even more methane, a potent GHG. The cold regions of the planet counteract the warmer parts of the planet and keep things from getting too warm. Ice also stores a great deal of water, keeping sea levels relatively stable and our coastal cities above water. Finally, the ice packs are the home of many plants and animals.

However, all this is very rapidly changing. Ice caps and glaciers are melting fast (Figure 9.4), and sea levels are rising as a result. The Greenland and Antarctic ice sheets are now losing more than three times as much ice a year than they were 30 years ago. Greenland's average annual melt from 2017 to 2020 was 20 percent more a year than at the beginning of the decade and more than seven times higher than its annual shrinkage in the early 1990s. Like most other places, the polar regions are rapidly warming. Some of the species living in polar areas will go extinct while others will move or otherwise adapt. Cold water fish that we depend on for our fish sticks will likely disappear. Polar bears will exist only in zoos.

But it may get much worse. A 2023 study[6] found that the Gulf Stream current in the Atlantic Ocean might collapse by 2050, perhaps as early as 2025. The cause of this is the increasing influx of cold freshwater from the melting ice sheets in Greenland entering the North Atlantic and disrupting the warm Gulf Stream. This would lead to catastrophic impacts. Northern Europe would significantly cool and initiate drought conditions, while intensifying Atlantic storms would affect the Caribbean and southern North America. Even Central Asia could be dried up since it depends heavily on moisture blown from the Atlantic by westerly winds. This general scenario was the plot of the 2004 Hollywood movie "The Day After Tomorrow." Bad times ahead.

FIGURE 9.4   Boulder Glacier, Washington State, 1932, TJ Hileman, GNP Archives –
1988, J DeSanto, U of M Library. Public Domain (https://www.usgs.gov/
media/images/boulder-ice-cave-glacier-1932-and-1988)

Up and Coming

Another major consequence of climate change is sea level rise.[7] The increasing
heat alone is causing the water in the oceans to expand, resulting in a sea level
rise of several feet. Miami is already flooding at high tide. In Italy, Venice is also

flooding, and the multi-billion dollar attempt to control such flooding has had serious issues. The Netherlands, some of which is below sea level, is threatened. Jakarta, the capital of Indonesia, is flooding due to sea level rise and the country is spending tens of billions of dollars to build a new capital at a higher elevation. Some low-lying cities are building or planning sea walls. The several feet of sea level rise due just to warming means that storm surges will be bigger and more devastating.

In addition to sea level rise just from warmer water, the melting ice in the Arctic, Greenland, and Antarctica is adding to the problem. Sea level rise is in the process of inundating some islands, including entire island nations such as Tuvalu in the South Pacific and the Maldives in the Indian Ocean. Higher islands are taking measures; the government of the island nation of Tonga is already in the process of moving settlements and agricultural fields away from the shores. Other island nations and low-lying coastlines face the same issues.

If all the world's ice melted, sea levels would rise some 100 meters (330 feet) and inundate coastlines worldwide, flooding considerable farmland and many coastal cities. In the US considerable portions of the southeast would be inundated, and Florida would essentially disappear (Figure 9.5). This would disrupt millions of people and result in the loss of considerable farmland and forests, not to mention trillions of dollars in property damage. In November of 2023, it was estimated that if an increase of 2°C becomes the new constant Earth temperature, sea levels will rise at least 40 feet.[8]

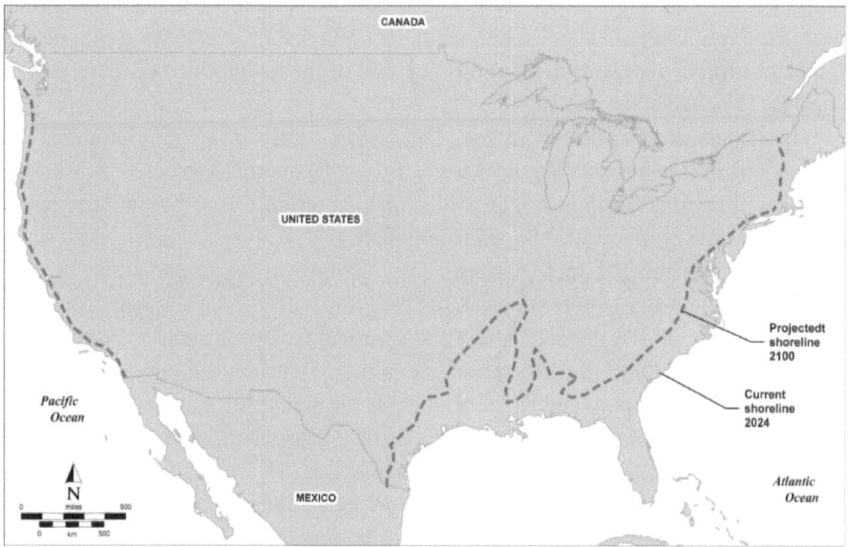

FIGURE 9.5 Map of the shoreline of the United States in 2100 due to projected sea level rise.

Fish Food

Alan Villiers, a former sailing ship captain, served on the last sailing ship to fish for cod on the Grand Banks. It was a Portuguese vessel. The crew were permanently drenched in near-freezing water, managed rigging under appalling conditions of wind, ice, and storm, hauled huge nets, and dealt with large flapping cod. They survived on bread, fresh cod, and endless, strong coffee. It is an amazing narrative of human survival under horrible conditions. Among countless excellent ethnographies of fishing, Alan Villiers' book *The Quest of the Schooner Argus* stands out.[9]

This was fishing in the old days. Even ENA, studying fisheries development in East Asia in the 1960s and 1970s, went through a typhoon on a small boat (safely in a typhoon shelter) and confronted storms and winds on open water. Fisheries work on Asian waterfronts in times past was always entertaining—one rubbed elbows with smugglers, pirates, and waterfront gangsters, some of whom had killed rivals.[10] Fishing in the 21st century is only somewhat better. Open-sea fishing rivals mining as the most dangerous occupation. Sadder still, though, it may not last much longer.

Ian Harrison and colleagues report: "Populations of freshwater species have declined by an average of 83 percent since 1970...freshwater fish extinction rates in the U.S. and Europe have been estimated to be more than 100 times their natural rates. Meanwhile, wetland loss is three times as high as forest loss."[11]

Fisheries expert Boris Worm and his colleagues have shown that, at current rates of fishing, worldwide commercial fishing will come to an end by 2050.[12] There will be no more wild fish in adequate quantities to support commercial efforts. International treaties, commissions, and protected areas may save a few fisheries, but the glory days are gone. World wild fish production peaked around 2018 and has been declining since. The world impact of declining fisheries is huge, going beyond issues of protein supply and economic welfare to the health of the oceans and the survival of ocean ecosystems.[13]

The reason is not difficult to see. The classic "tragedy of the commons" has always affected fisheries.[14] Ancient writers recount overfishing cases. Stories and folktales from the Northwest Coast of North America to the South Seas record famines caused by people taking too many fish.

Fish are notoriously hard to manage. They not only move around all the time but also do it in the water, where we can't see them the way we can see migrating herds of bison. They tend to appear in vast shoals, then disappear again, sometimes for years. Even when they are predictable, like salmon, runs may vary greatly from year to year. Populations may suddenly explode or disappear for reasons often obscure until scientists take a very long look. Overfishing makes stocks more variable—so they may look as if they are coming back due to random fluctuations.[15]

Even so, the effects of overfishing are perfectly clear after a while. Increased catching pressure means decreasing fish populations. Unfortunately, the ability of fishermen to deny the obvious is almost infinite. Anyone with experience in fishing has heard over and over, "The fish just aren't biting these days," "the fish have just gone away for a while," "it won't hurt to take just a few more," and so on and on and on. The reality is exactly what Hardin identified: the temptation to take one more from a common resource when there is no enforcement is hard to resist, and those who do resist tend to quit the fishery, leaving it to the least responsible.

Some of the slack has been picked up by fish farming (aquaculture)—penning fish with nets in sea or lake waters—but this allows too many escapes. There are, for instance, millions of escaped salmon in Chile, British Columbia, and elsewhere, devastating native fish.[16] Pollution caused by fish farms is usually treated as an "externality." It is simply flushed out with the water and becomes someone else's problem. In other words, the rivers and oceans are treated as a common-pool resource for the purpose of waste disposal. Privatization that does not specify such problems for the private farmer is not true privatization; it is exploitation of the public via abuse of rivers and oceans. Such pollution can be controlled, of course, by government laws, but, once again, politicians are corruptible. Further, much of the feed for these farmed fish is often small fish caught wild at sea. This is a terrible idea, depleting vast natural fisheries to feed a much smaller number of fish, often for luxury diners.

China, by far the world's leading fish-farming nation, has long been warned against this,[17] and indeed, China's traditional system is much less damaging. China accounts for well over 50 percent of all farmed fish production, now almost four times China's wild catch.[18] China has very old pond stocking systems that dispense with the need for wild-caught food fish. China's native carp species (including the common carp, now gone worldwide) can live off ordinary freshwater resources. The grass carp lives on grass and other vegetation, which can be raised in fishponds and on dykes. Other carp live on small invertebrates in the water or mud. Productivity is extremely high. It does not depend on food inputs other than some fertilizing of the water.[19]

Relatively neglected resources like carp deserve more attention. In the US, the fish is despised, partly because it so often eats algae that give it a muddy flavor. If carp avoid algae, they taste good but are bony. They flourish in waters too muddy and polluted for other fish, and thus have become identified with nonaffluent fishers and so have bad reputations.

Other types of marine resources can also be farmed. Seaweeds and seagrass are now cultivated. Shellfish are cultivated on an enormous scale. Mussels, clams, and oysters have been cultivated for centuries, but now their production is booming.[20] For them, pollution and destruction of bedding areas are the only limits. Shrimp are also farmed. The supply of wild shrimp could not keep up with the demand, so commercial shrimp farming began in the 1970s, and most of the shrimp now consumed comes from farms. However, the shrimp in these farms are not genetically diverse and so are susceptible to diseases, which caused a number of farms to have to shut down. These outbreaks, combined with other issues, resulted in more government regulation and a stabilization in supply.

The problem with regulation is that it is difficult to convince governments to do it. Legislators usually have other priorities than saving fish and prefer to keep fishermen and fish consumers happy (and voting for the politicians) by allowing them to have their way. This destroys the fishery in short order, but the politicians rarely think beyond the next election or coup, so they fail to plan for the long haul. Problems include such simple matters as size limits. These often were invoked to preserve young fish, but by targeting older fish, they often lead to catching the most successful reproducers, selecting for smaller fish overall, and unbalancing the aquatic ecosystem.[21]

A misguided attempt to control overfishing was to restrict entry to fisheries—privatizing access rather than the resource. The commonest way to do this is to limit the number of boats allowed into a fishing area, each with "individual transferable quotas" (ITQs).[22] Individual boats had to pay high fees and had quotas theoretically set by management experts. The problems were legion, but largely boiled down to the obvious one: restricting entry was not the same as direct restriction of take. Too many boats got into the fishery. The price of licenses skyrocketed, shutting out less affluent fishers. Since, in the old days, fishing was an easy job to enter, members of minority groups very often resorted to it. They were hard hit by the ITQ movement. Boats of affluent fishers increased catching power and took more than their share. Some boats became monstrosities, similar to millionaires' yachts, simply because they could catch enormous amounts in short seasons with ITQs (ENA, personal research). Evelyn Pinkerton has emerged as a leading critic of ITQ's and, above all, a proposer and chronicler of alternatives in community-based fishery.[23]

Another attempt was to shorten the fishing season. A combination of enormously increased catching power—huge boats with enormous nets—and collapse of fish populations led to an extreme situation in British Columbia: during one year, the herring season was limited to 40 minutes (not a typo, 40 *minutes*). Herring populations continued to decline anyway, leading, among other things, to a collapse of seabird populations.[24] It makes more sense to set a limit to the catch and let fishermen take their time getting it.

Regulators need better estimates of fish populations and what they can bear. Since the 1960s, the gold standard has been "maximum sustainable yield" (famous as MSY), but bitter experience has shown that governments and other regulators are hopelessly overoptimistic—and frequently downright dishonest—about that figure. The US law now sets limits according to calculations of "optimum yield," usually lower than MSY. Overfished stocks are supposed to be allowed to recover. This has been spottily applied.

If protecting marine resources is taken seriously, it works. Elephant seals have increased from near extinction, from the 20 Northern Elephant Seals known in 1880 to hundreds of thousands now. Humpback whales, but sadly not northern right whales, have increased greatly.[25] These are large, charismatic species that attract public attention. Less evident fish are in much more trouble.

The best regulatory method is the creation of reserves completely closed to fishing. This was historically done in several cases. There were religiously tabooed waters in some areas, especially where Buddhism had introduced the idea of protecting all animals. There were waters reserved for royalty, nobility, or local residents. There were temporary closures of overfished waters. More recently, large areas of ocean have been closed as absolute sanctuaries. These serve as breeding areas for endangered or depleted fish. Since fish naturally move around, these areas provide source areas for fish that replenish neighboring waters that are still heavily fished. It has been estimated that protecting five percent of the ocean would increase fish catches for food by 20 percent.[26]

Another worthwhile effort is getting restaurants and markets to label fish according to sustainability of production. This is undercut by the rampant tendency of sellers at all levels to mislabel fish. Sometimes this is simple ignorance, or local folk usage, or assumption that the public do not know that an obscure fish is good (and thus give it a more saleable name), but usually it is malicious. Cheap, inferior fish are labeled with choice names. DNA testing of fish in markets is now a hobby and growth industry for students who want to assess their genetics knowledge while doing public service. Usually the mislabeling is fairly benign, like listing various red fish as "red snapper," but sometimes it is dangerous. Buyers are sometimes misled, for instance, into buying escolar, a fish that contains a waxy fat that causes violent diarrhea.[27]

Conservation sometimes works and sometimes fails. Populations of North Atlantic cod, commercially fished for more than a thousand years, crashed in the early 1990s due to overfishing.[28] In some areas, such as the Grand Banks off New England, cod populations completely collapsed, and fishing was prohibited (and remains so) in the hope that the populations would recover. Elsewhere in the North Atlantic, cod populations have rebounded, but not in the Grand Banks, although a very slight improvement has been recorded. The North Atlantic fishery was already substantially depleted by 1500. After that, exploitation skyrocketed, wiping out everything from local whale populations to oyster beds.[29]

In the North Pacific, the annual catch of Pacific cod is regulated to prevent a catastrophic collapse. The annual catch of King crab in the North Atlantic is also regulated due to a precipitous drop in the populations in the early 1980s. Many other fisheries are now regulated, including wild salmon, and various shellfish, including abalone. Unfortunately, this sometimes comes too late. Abalone, depleted and subject to natural predation, does not recover easily.[30] The population of snow crabs in the North Pacific has crashed and may not recover.[31] The issue seems to be increasingly warm water, impacting the food supply of the crabs.

In these and many other cases, not only did the tragedy of the commons run a devastating course, but it also did so with government help. In Malaysia and British Columbia, governments actively intervened to prevent protection of local areas. The pressure from fishing interests, who could deliver votes and funds, was stronger than the pressure from conservationists.

A depressing insight into this cynical and cold process is the continual failure of the nations around the North Sea to protect the herring there. This is no new thing; herring, along with sturgeon, eels, and other popular fish, were overfished locally by the end of the Middle Ages.[32] The modern governments of the region are the most responsible, welfare-oriented, and reliable in the world. They include Scandinavian countries, the United Kingdom, and the Low Countries. They are, in many ways, ideal states, regularly topping the world surveys of "happiest" and "most democratic" nations. They also once depended on herring as a cheap protein resource for their people. They know perfectly well that their herring are extremely overfished. The two World Wars forced closure of the fishery, so during and after both wars, the herring bounced back with astonishing speed. Nothing was done about conservation after the wars, however, and the herring stock promptly crashed again.[33]

The EU was formed in part to allow regional conservation and environmental protection, but it has done nothing effective about the herring. One would wonder about the sanity of the ruling classes. Most of them simply do not care. They want to please shortsighted fishers who vote and are willing to let a billion-dollar resource go. In general, the EU has not taken fish seriously.

The fate of herring in the North Sea taught fisheries regulators exactly nothing. Archaeology shows that Native peoples of British Columbia managed herring sustainably for 10,000 years, only to see the populations decline by over 90 percent in the last few hundred years. The fate of herring worldwide is an extreme case of human irrationality. The belief that people act in their rational self-interest must deal with the fact that people may be "rational" in the very short run by being suicidal in the long run. Governments take over control but do not take on the responsibilities.[34]

Another destructive approach is trawling—dragging a huge net over the sea floor—to scoop up whatever is there, leaving a barren track in its wake. Nobody even pretends that trawling makes economic sense, and nobody denies it is the most destructive and wasteful common type of fishing, yet trawling is heavily subsidized and protected by the EU (even the sober scientific journal *Nature* had some scathing editorial comments on this).[35] It even goes on in supposedly protected areas through blatant crime and corruption.

The problem of subsidizing fisheries is not limited to the EU, as most national fisheries worldwide are heavily subsidized.[36] Fisherfolk tend to be far from affluent as their operating costs are high. In democracies, they vote and can dominate waterfront districts. The result is that, worldwide, more money is spent catching fish than the fish are worth. It costs about $1.50 to harvest $1 worth of fish— substantially more in some countries. Thus, fishing goes far beyond a level that is economically viable. Diminishing returns that should shut down a fishery fail to do so. China has recently dealt with the problem by cutting subsidies for fishing and actually subsidizing fishermen to retire and take their boats out of service. China's enormous population of fishermen is aging, and many are happy to take this step.[37] There is a long way to go, however. China's vast fleet ravages the world's oceans, often working out treaties with weak governments to fish heavily in their waters.

In the southern oceans, fish and krill are being taken for industrial use or for luxury, despite regulations and recommendations and in spite of common sense.[38]

There, and everywhere, whales, seabirds, seals, and other life forms (such as turtles in the warmer regions) depend on these marine resources, and so are exterminated along with the fish. Whales used to consume 430 million tons of krill a year, far more than exists today. Their consumption and digestion cycled nutrients more rapidly, allowing the seas to be more productive.[39]

Some types of fish are especially targeted. Tuna is so popular in Japan that a single bluefin is worth hundreds of thousands, even millions, of dollars, making conservation almost hopeless. Shark fins (fishers often cut off the fins and throw the shark back into the water for a slow death) are similarly a delicacy in Chinese cuisine, leading to reductions of 98 percent or more in some shark populations. In highly populated, heavily fished areas that are not stringently conserved, the sharks are almost gone, replaced by rays. In 2009, Palau created a 600,000 square kilometer (the size of France) shark sanctuary providing some protection. Sturgeon are gone almost everywhere because of demand for caviar, but there is some hope, as they are now being farm-raised.[40]

Other types of fish are less exploited, but probably all significant stocks in the world today are declining. Underfished stocks are generally those that do not command a high price. They are often overfished anyway, however, as by-catch in hauls targeted at more profitable fish or as feed fish for the fish farms. Conversely, limits on overfishing preferred fish have the desirable spillover effect of saving fish that would otherwise have been caught.[41] Today's less preferred fish is tomorrow's desperately needed fish, once its more desirable neighbors have been fished out.

Still more irrational is the practice of taking tiny, immature fish for fertilizer or for feed for fish farms. Taken in mass by trawlers, they bring a few dollars per ton. Many or most of these tiny fish would have grown up to be commercially valuable species worth many dollars per pound.[42]

Roving fishing fleets from heavily fish-dependent nations are currently sweeping the world, often ruining small nations by moving in, taking their fish illegally or after corrupting politicians, and then moving on from the ruin. Such fleets underreport their catch. This piracy has been analyzed by fishery experts Fikret Berkes and Boris Worm.[43] China reports fairly honestly on the notorious fishing it carries out in its own waters but underreports its international catch.

Meanwhile, climate change is warming the oceans more and more rapidly, moving fish populations north. Although some fish increase in numbers with the move, it usually has negative effects since not all marine resources can move fast and since warm water holds less oxygen and thus allows less growth. For instance, northern New England's lobster populations have expanded, while southern ones have declined.[44]

A final insult is that the world's rivers, lakes, and seas are often considered more valuable as power sources, or even as garbage dumps, than as fish suppliers. Hydroelectric power generated by big dams has led to the destruction of freshwater fisheries in countless rivers. Mining wastes, urban wastes, farm wastes, and chemical wastes have destroyed the fisheries of countless more. The proposed Pebble mine in Alaska would wipe out the 40 million sockeye salmon of the Bristol Bay fishery just to produce a small amount of gold. This is economically irrational but popular with local politicians who get campaign contributions from mining interests. In early 2023, the permit for the mine was denied, but that decision is now under appeal.

By comparison, the Columbia River fishery, once much larger than Bristol Bay's, is down to 3.5 million fish.[45] It was sacrificed largely to hydroelectric power and irrigation; huge dams blocked the river, with inadequate provisions for salmon and other fish that formerly ran up the river to spawn. Often, the problem could be solved by fish ladders or the equivalent, but that would contribute slightly to the cost of the dam. So, a multimillion-dollar fishery is sacrificed to save a few extra dollars on the dam. This pays off because no one is charged for the ruin of the fishery—even though thousands of people suffer from loss of their fisheries.

Cleanup of dammed rivers is rare and imperfect. At least, there is a movement to take out unnecessary or superannuated dams to restore the fish. The Elwha River in Washington state suffered from two old and no longer useful dams that blocked salmon migration. These dams have been removed, whereupon the fishery began recovery with amazing speed. Many other dams worldwide are being removed, with similar success.[46]

Unfortunately, new dam construction goes on at a much more rapid pace than the removal of old dams.[47] The huge and extremely fish-rich Mekong River in Asia is being turned into a succession of reservoirs, with consequent ongoing destruction of one of the greatest protein resources in the world—a fishery that provided high-quality protein to tens of millions of people. The power generated by the dams goes in large part to well-to-do urbanites.

Traditional Regulation, Contemporary Optimism

Indigenous peoples generally manage their fisheries more efficiently. For example, Native people in the Northwest Coast of North America, from northern Oregon to northern Alaska, depended largely on fish for food. Over thousands of years, they devised a practical strategy for managing stocks. Until about 200 years ago, salmon and related fish abounded in even the tiniest streams, though a single fisher with a net could "rob a creek" (catch all the fish, leaving the creek lifeless) in a few days. It took genuine forbearance and a strong conscience to avoid doing this.

The basis of this management was respect and responsibility. By far the most important behavior involved was taking small amounts of a resource—only what was needed for immediate consumption and for storage for the year. There was more, however. Native people cleaned up streams, improved spawning areas, and removed barriers. At least on Vancouver Island, they stocked small streams that lacked fish. Many groups returned processed wastes, at least the bones, to the water, thus providing nutrients for developing young. Salmon die after spawning so that the growing fry can get the nutrients of the parental bodies—the ultimate sacrifice.

More important was the ideology behind this approach. Northwest Coast people assumed fish were intelligent beings similar in mentality to humans. Many groups believed that the fish lived in undersea houses, where they could become humans or humanlike. They put on their fish skins to sacrifice themselves for their human relatives. This meant that fish were kin, deserving of respect and consideration. Even groups less committed to humanizing the fish treated them with respect. That meant not killing them unnecessarily, using all the fish when it was taken, and not taking enough fish to diminish the total population. The same rules applied to game animals and birds, but fish were the most important resource and thus treated with special consideration. These principles were embedded in a wider worldview of unity with the surrounding world. Trees, rocks, mountains, and streams had their spirits, as did fish. Even if these beings were not taken as literal persons, they were important parts of the world, deserving of all respect.[48] Respect for the resource, instilled in religion and tradition, counteracts a tragedy of the commons result.

Many other traditional people worldwide have similar ideas and rules. In fact, almost every group of people in a small, traditional society that depends on fish has developed such rules. They would not have lasted long enough to become "traditional" if they had not. For this to work, people must be emotionally connected to the fish, at least loving them as food, at most loving them as sacred beings or literal kin. Without these emotional and moral ties, people almost invariably succumb to "tragedy of the commons" logic and wipe out the resource.[49]

However, it is not necessary to believe in the personhood of fish to conserve them. The lobster fishers of Maine have conserved and protected lobster populations for a couple of centuries. They manage by having a tightly organized fishery where everyone knows everyone else—or at least has direct or indirect ties. They can monitor catching methods and amounts caught. They can quickly identify anyone not following rules, and public opinion is generally enough to bring such people into line. At worst, they can be kept from getting further licenses. A classic study of this system was done by James Acheson.[50]

Stig Gezelius[51] usefully compared three ways of enforcing fisheries rules: Hobbesian (tight regulation), Habermasian (rational dialogue), and Durkheimian (social pressure) based on examples in fishing villages in the North Atlantic. The third worked best. The implications obviously extend far beyond fisheries.

On the plus side, a landmark international treaty for oceans was signed by most world nations on March 4, 2023. It aims for protection of 30 percent of oceans by 2030 (1.2 percent is now protected), sharing newly discovered genetic resources with all countries, regulating seabed mining, and related goals. It still must be ratified and implemented. From improving farmed fish to saving wild ones to signing comprehensive treaties to save the world's oceans, possibilities for ending the decline of world fisheries are many and various. Only the will of governments is required.[52]

A New Hope?

Since we are quite dependent on the ocean, it is obvious that efforts must be made to improve its health. We have already discussed the major issue of depleted fisheries and pollutants, but other important issues include the loss of coral and plankton. Various non-governmental organizations, such as the Ocean Conservancy, The Ocean Foundation, The Nature Conservancy, Greenpeace, and many others, are involved in trying to protect the oceans.

Coral reefs are important for several reasons. They provide protection from storms and help reduce coastal erosion. Many of the islands in the tropical Pacific Ocean are made of, and protected by, coral. The marine life associated with coral reefs provides a source of food for coastal communities. Finally, reefs are tourist destinations and generate employment and economic growth.

Coral reefs are under threat in many areas, due mostly to increasing water temperatures but also to coral mining, pollution (e.g., oil, agricultural chemicals, and sunscreen), water acidification, water deoxygenation, disease, and mechanical disturbances such as underwater construction (e.g., for harbor access) and damage from ships. Among the most endangered reef systems are the Great Barrier Reef in Australia and the various reefs in Indonesia, the Caribbean, and elsewhere.

There are various efforts and projects underway to mitigate the loss, and restoration of, coral reefs.[53] Methods of restoration include transplanting nursery-grown coral onto existing reefs, relocating healthy coral to stressed reefs, making coral substrate using mineral accretions or to encourage the growth of new coral by installing things like small ceramic plates for the coral to attach themselves to, and to provide habitat for other reef species, and the construction of artificial reefs, such as with purposefully sunken (but cleaned first) ships. Inhibiting these efforts are the increasing water temperature and acidification.

Plankton is a general category that includes phytoplankton (plants) and zooplankton (animals) and is a major food source for many fish, shellfish, and some whales. Phytoplankton produce some 50 percent of the world's oxygen and store substantial amounts of carbon. The impact of climate change on plankton, especially phytoplankton, is not well understood, but some fear that the amount of phytoplankton is decreasing (others think it is increasing).

As phytoplankton store carbon, an increase in their mass would help offset GHG emissions. Thus, there is an idea to "fertilize" the oceans with iron to encourage the growth of phytoplankton, but this may turn out to be impractical.

What Can Individual People Do?

There are many things that individuals can do to help improve the health of the oceans. In some cases, people are making lemonade from the problem lemons. For example, sargassum (macroalgae) that has washed up on beaches in Mexico creates a nuisance when it decays. To put that mess to use, a local man formed a company, Sargablock, and is turning the sargassum into bricks for use in construction.[54]

Many other individual actions could be taken. Use less plastic and support trash cleanup on beaches and waterways. Eat sustainable seafoods. Enjoy, but do not damage coral reefs. Avoid ocean-polluting materials such as cleaners. Educate yourself on the issues. Finally, and most importantly, elect politicians that care about the planet.

Notes

1 Helvarg, David. 2023. "Florida Waters Hit 101. What Does That Mean for the Ocean?" *Los Angeles Times*, Aug. 2, A11.
2 Smith, Jesse, and Sacha Vignieri. 2021. "A Devil's Bargain." *Science* 373:34–35; Pinheiro, Hudson T., et al. 2023. "Plastic Pollution on the World's Coral Reefs." *Nature* 619:311–322.
3 Wilcox, Chris, Erik Van Sebille, and Britta Denise Hardesty. 2015. "Threat of Plastic Pollution to Seabirds Is Global, Pervasive, and Increasing." *Proceedings of the National Academy of Sciences* 112:11899–11904.
4 Smith, Jesse, and Sacha Vignieri. 2021. "A Devil's Bargain." *Science* 373:34–35; Subramanian, Meera. 2022. "Can Nations Rein in Plastics Pollution?" *Nature* 611:650–653.
5 Kwon, Diana. 2023. "Three Ways to Solve the Plastics Pollution Crisis." *Nature* 616:234–237. Quote from p. 234.
6 Ditlevsen, Peter, and Susanne Ditlevsen. 2023. "Warning of a Forthcoming Collapse of the Atlantic Meridional Overturning Circulation." *Nature Communications* 14:article 4254; also see Bell, Art, and Whitley Strieber. 2000. *The Coming Global Superstorm.* New York: Pocket Books.
7 Goodell, Jeff. 2017. *The Water Will Come: Rising Seas, Sinking Cities, and the Remaking of the Civilized World.* Boston, MA: Little, Brown and Company.
8 International Cryosphere Climate Initiative (ICCI). 2023. *State of the Cryosphere 2023 – Two Degrees Is Too High.* Stockholm, Sweden. 62 pp.
9 Villiers, Alan. 1951. *The Quest of the Schooner Argus.* New York: Charles Scribner's Sons.
10 Anderson, E. N. 1978. *Fishing in Troubled Waters. Taipei: Orient Cultural Service; 2007. Floating World Lost.* New Orleans: University Press of the South. Heffernan, Olive. 2024. *The High Seas: Ambition, Power and Greed on the Unclaimed Ocean.* New York: Profile Books.
11 Harrison, Ian, Robin Abell, William Darwall, Michele L. Thieme, David Tickner, and Ingrid Timboe. 2018. "The Freshwater Biodiversity Crisis." *Science* 362:1369.
12 Worm, Boris. 2016. "Averting a Global Fisheries Disaster." *Proceedings of the National Academy of Sciences* 113:4895–4897. Worm, Boris, Edward B. Barbier, Nicola Beaumont, J. Emmett Duffy, Carl Folke, Benjamin S. Halpern, Jeremy B. C. Jackson, Heike K. Lotze, Fiorenza Micheli, Stephen R. Palumbi, Enric Sala, Kimberley A. Selkoe, John J. Stachowicz, and Reg Watson. 2006. "Impacts of Biodiversity Loss on Ocean Ecosystem Services." *Science* 314:787–790; for "peak fish," see World Almanac. 2023. *The World Almanac and Book of Facts 2023.* New York: World Almanac Books, p. 147.
13 Kroodsma, David A., Juan Mayorga, Timothy Hochberg, Nathan A. Miller, Kristina Boerder, Francesco Ferretti, Alex Wilson, Bjorn Bergman, Timothy D. White, Barbarra A. Block, Paul Woods, Brian Sullivan, Christopher Costello, and Boris Worm. 2018. "Tracking the Global Footprint of Fisheries." *Science* 359:904–908.
14 McEvoy, Arthur F. 1986. *The Fisherman's Problem: Ecology and Law in the California Fisheries, 1850–1980.* Cambridge: Cambridge University Press.
15 Anderson, Christian N. K., Chih-hao Hsieh, Stuart A. Sandin, Roger Hewitt, Anne Hollowed, John Bedington, Robert M. May, and George Sugihara. 2008. "Why Fishing Magnifies Fluctuations in Fish Abundance." *Nature* 452:835–839.

16 Figueroa-Muñoz, Guillermo, et al. 2022. "Co-management of Chile's Escaped Farmed Salmon." [Letter.] *Science* 378:1060–1061.

17 Cao, Ling, Rosamond Naylor, Patrik Henriksson, Duncan Leadbitter, Marc Metian, Max Troell, and Wenbo Zhang. 2015. "China's Aquaculture and the World's Wild Fisheries." *Science* 347:133–134.

18 World Almanac. 2023. *The World Almanac and Book of Facts 2023*. New York: World Almanac Books.

19 Wen, Dazhong and David Pimentel. 1990. "Energy Flow in Agroecosyste in Northeast China." In *Agroecology: Researching the Ecological Basis for Sustainable Agriculture*, Stephen Gliessman (ed.), pp. 322–336. New York: Springer-Verlag. Also ENA, personal research.

20 Naylor, Rosamond L., et al. 2021. "A 20-Year Retrospective Review of Global Aquaculture." *Nature* 591:551–563.

21 Barneche, Diego R., D. Ross Robertson, Craig R. White, and Dustin J. Marshall. 2018. "Fish Reproductive-Energy Output Increases Disproportionately with Body Size." *Science* 360:642–645; Nadasdy, Paul. 2004. *Hunters and Bureaucrats: Power, Knowledge, and Aboriginal-State Relations in the Southwest Yukon*. Vancouver: University of British Columbia Press.

22 Finley, Carmel. 2011. *All the Fish in the Sea: Maximum Sustainable Yield and the Failure of Fisheries Management*. Chicago: University of Chicago Press; Finley, Carmel. 2017. *All the Boats on the Ocean: How Government Subsidies Led to Global Overfishing*. Chicago: University of Chicago Press.

23 Pinkerton, Evelyn. 2015. "The Role of Moral Economy in Two British Columbia Fisheries: Confronting Neoliberal Policies." *Marine Policy* 61:410–419; Pinkerton, Evelyn. 2017. "Hegemony and Resistance, Disturbing Patterns and Hopeful Signs in the Impact of Neoliberal Policies on Small-Scale Fisheries around the World." *Marine Policy* 80: 1–9; Pinkerton, Evelyn (ed.). 1989. *Cooperative Management of Local Fisheries: New Directions for Improved Management and Community Development*. Vancouver: University of British Columbia; Pinkerton, Evelyn, and Martyn Weinstein. 1995. *Fisheries That Work: Sustainability through Community-Based Management*. Vancouver: David Suzuki Foundation.

24 Birkenbach, Anna M., David J. Kaczan, and Martin D. Smith. 2017. "Catch Shares Slow the Race to Fish." *Nature* 544:223–226; Also ENA, personal research in British Columbia, 1984–1985.

25 Duarte, Carlos M., Susana Agusti, Edwar Barbier, Gregory L. Britten, Juan Carlos Castilla, Jen-Pierre Gattuso, Robinson W. Fulweiler, Terry P. Hughes, Nancy Knwolton, Catherine E. Lovelock, Heike K. Lotze, Milica Predragovic, Elvira Poloczanska, Callum Roberts and Boris Worm. 2020. "Rebuilding Marine Life." *Nature* 580:39–51.

26 Cabral, Reniel B., et al. 2020. "A Global Network of Marine Protected Areas for Food." *Proceedings of the National Academy of Sciences* 117:28134–28139.

27 Kroetz, Kailin, et al. 2020. "Consequences of Seafood Mislabeling for Marine Populations and Fisheries Management." *Proceedings of the National Academy of Sciences* 117:30318–30323.

28 Bavington, Dean L. 2010. *Managed Annihilation: An Unnatural History of the Newfoundland Cod Collapse*. Vancouver: University of British Columbia Press. Also see Kurlansky, Mark 1998. *Cod: A Biography of the Fish that Changed the World*. Westminster, MD: Penguin Books.

29 Holm, Poul, et al. 2022. "The North Atlantic Fish Revolution (ca. AD 1500)." *Quaternary Research* 108:92–106.

30 Menzies, Charles R. 2016. *People of the Saltwater: An Ethnography of Git lax m'oon*. Lincoln: University of Nebraska Press, pp. 119–130, provides a stunning example of Indigenous good management vs. Settler wipeout.

31 Szuwalski, Cody S., et al. 2023. "The Collapse of Eastern Bering Sea Snow Crab." *Science* 382:306–310.

32 Hoffmann, Richard C. 2023. *The Catch: An Environmental History of Medieval European Fisheries*. Cambridge: Cambridge University Press.
33 For a history of herring see Thornton, Thomas, and Madonna L. Moss. 2021. *Herring and People of the North Pacific: Sustaining a Keystone Species*. Seattle: University of Washington Press.
34 Cressey, Daniel. 2013. "EU Fishing Vote Foments Anger." *Nature* 504:341; Dureuil, Manuel, Kristina Boerder, Kirsti A. Burnett, Rainer Froese, and Boris Worm. 2018. "Elevated Trawling Inside Protected Areas Undermines Conservation Outcomes in a Global Fishing Hot Spot." *Science* 362:1403–1407; Froese, Rainer. 2011. "Fishery Reform Slips through the Net." *Nature* 475:7; Pitcher, C. Roland, et al. 2022. "Trawl Impacts on the Relative Status of Biotic Communities of Seabed Sedimentary Habitats in 24 Regions Worldwide." *Proceedings of the National Academy of Sciences* 119:2019449119.
35 Nature. 2023. "The Hypocrisy Threatening the World's Oceans." (Editorial.) *Nature* 621:7.
36 Finley, Carmel. 2017. *All the Boats on the Ocean: How Government Subsidies Led to Global Overfishing*. Chicago: University of Chicago Press.
37 Wang, Kaiwen, Matthew N. Reimer, and James E. Wilen. 2023. "Fisheries Subsidies Reform in China." *Proceedings of the National Academy of Sciences* 120:e2300688120.
38 Brooks, Cassandra M., et al. 2022. "Protect Global Values of the Southern Ocean Ecosystem." *Science* 378:477–479; Burgess, Matthew G., et al. 2018. "Protecting Marine Mammals, Turtles, and Birds by Rebuilding Global Fisheries." *Science* 359:1255–1258; Savoca, Victor, et al. 2021. "Baleen Whale Prey Consumption Based on High-Resolution Foraging Measurements." *Nature* 599:85–90.
39 Whales were destroyed early, largely for their oil. A New York court in the early 19th century ruled they were fish, not mammals, in spite of the scientists—because fish oil was taxed and mammal oil was not (Burnett, D. Graham. 2007. *Trying Leviathan*. Princeton, NJ: Princeton University Press.). Exploitation trumps science.
40 MacNeil, M. Aaron, et al. 2020. "Global Status and Conservation Potential of Reef Sharks." *Nature* 583:801–806; Madigan, Daniel J., Andre Boustany, and Bruce B. Collette. 2017. "East Not Least for Pacific Bluefin Tuna." *Science* 357:356–357; Pacoureau, Nathan, et al. 2021. "Half a Century of Global Decline in Oceanic Sharks and Rays." *Nature* 589:567–571; Pala, Christopher. 2005. "Ban on Beluga Caviar Points to Sturgeon's Worldwide Decline." *Science* 310:37; Simpfendorfer, Colin A., et al. 2023. "Widespread Diversity Deficits of Coral Reef Sharks and Rays." *Science* 380:1155–1160; Telesca, Jennifer. 2021. *Red Gold: The Managed Extinction of the Giant Bluefin Tuna*. Minneapolis: University of Minnesota Press.
41 Oremus Kimbery L., Eyal G. Frank, Jesse Jian Adelman, Seleni Cruz, Janna Herndon, Brad Sweell, and Lisa Suatoni. 2023. "Underfished or Unwanted?" *Science* 380:585–588.
42 Anderson, E. N. 1978. *Fishing in Troubled Waters. Taipei: Orient Cultural Service; 2007. Floating World Lost*. New Orleans: University Press of the South; Cao, Ling, Rosamond Naylor, Patrik Henriksson, Duncan Leadbitter, Marc Metian, Max Troell, and Wenbo Zhang. 2015. "China's Aquaculture and the World's Wild Fisheries." *Science* 347:133–134; Sumaila, U. Rashid, W. W. L. Cheung, et al. 2021. *Sink or Swim: The Future of Fisheries in the East and South China Seas*. Hong Kong: ADM Capital Foundation.
43 Berkes, F., T. P. Hughes, R. S. Steneck, J. A. Wilson, D. R. Bellwood, B. Crona, C. Folke, L. H. Gunderson, H. M. Leslie, J. Norberg, M. Nyström, P. Olsson, H. Österblom, M. Scheffer, and B. Worm. 2006. "Globalization, Roving Bandits, and Marine Resources." *Science* 311:1557–1558.
44 Cheung, William W. L., Reg Watson, and Daniel Pauly. 2013. "Signature of Ocean Warming in Global Fisheries Catch." *Nature* 497:365–368; Moore, J. Keith, et al. 2018. "Sustained Climate Warming Drives Declining Marine Biological Productivity." *Science* 359:1139–1143. For the lobster story, Le Bris, Arnault, et al. 2018. "Climate

Vulnerability and Resilience in the Most Valuable North American Fishery." *Proceedings of the National Academy of Sciences* 115:1831–1836.

45 Cornwall, Warren. 2019. "Minefield." *Science* 365:1366–1372.

46 Lohan, Tara. 2018. "The Elwha's Living Laboratory: Lessons from the Largest Dam-Removal Project." *The Revelator*, Oct. 1; O'Connor, J. E., J. J. Duda, and G. E. Grant. 2015. "1000 Dams Down and Counting." *Science* 348:496–497.

47 Barbarossa, Valerio, et al. 2020. "Impacts of Current and Future Large Dams on Geographic Range Connectivity of Freshwater Fishes Worldwide." *Proceedings of the National Academy of Sciences* 117:3648–3655.

48 Anderson, E. N., and Raymond Pierotti. 2022. *Respect and Responsibility in Pacific Coast Indigenous Nations: The World Raven Makes*. Cham, Switzerland: SpringerNature. There are several outstanding books by Indigenous Northwest Coast writers, e.g., Atleo, E. Richard. 2004. *Tsawalk: A Nuučaańuł Worldview*. Vancouver: University of British Columbia Press, and Atleo, E. Richard. 2011. *Principles of Tsawalk: An Indigenous Approach to Global Crisis*. Vancouver: University of British Columbia Press; Coté, Charlotte. 2010. *Spirits of Our Whaling Ancestors: Revitalizing Makah and Nuučaańuł Traditions*. Seattle: University of Washington Press; Menzies, Charles R. 2016. *People of the Saltwater: An Ethnography of Git lax m'oon*. Lincoln: University of Nebraska Press, pp. 119–130.

49 Bailey, Kevin M. 2018. *Fishing Lessons: Artisanal Fisheries and the Future of Our Oceans*. Chicago: University of Chicago Press; Cordell, John (ed.). 1989. *A Sea of Small Boats*. Cambridge, MA: Cultural Survival; Dyer, Christopher, and James R. McGoodwin (eds.). 1994. *Folk Management in the World's Fisheries: Lessons for Modern Fisheries Management*. Niwot: University Press of Colorado; Johannes, R. E. 1981. *Words of the Lagoon: Fishing and Marine Lore in the Palau District of Micronesia*. Berkeley: University of California Press; Ruddle, Kenneth, and R. E. Johannes, eds. 1983. *The Traditional Knowledge and Management of Coastal Systems in Asia and the Pacific*. Jakarta: UNESCO, Regional Office for Science and Technology for Southeast Asia; Svanberg, Ingvar, and Alison Locker. 2020. "Ethnoichthyology of Freshwater Fish in Europe: A Review of Vanishing Traditional Fisheries and Their Cultural Significance in Changing Landscapes from the Later Medieval Period with a Focus on Northern Europe." *Journal of Ethnobiology and Ethnomedicine* 16:68, doi.org/10.1186/s13002-020-00410-3.

50 Acheson, James M. 1987. "The Lobster Fiefs Revisited: Economic and Ecological effects of Territoriality in Maine Lobster Fishing." In *The Question of the Commons*, Bonnie McCay and James Acheson (eds.), pp. 37–65. Tucson: University of Arizona Press; Acheson, James M. 1998. "Lobster Trap Limits: A Solution to a Communal Action Problem." *Human Organization* 57:43–62; Acheson, James M. 2006. "Lobster and Groundfish Management in the Gulf of Maine: A Rational Choice Perspective." *Human Organization* 65:240–252.

51 Gezelius, Stig S. 2007. "Three Paths from Law Enforcement to Compliance: Cases from the Fisheries." *Human Organization* 66:414–425.

52 For further information on the history of fisheries, see Dyson, John. 1977. *Business in Great Water*. London: Angus and Robertson; McCay, Bonnie. 1998. *Oyster Wars and the Public Trust: Property Law, and Ecology in New Jersey History*. Tucson: University of Arizona Press.

53 Boström-Einarsson, L., R. C. Babcock, E. Bayraktarov, D. Ceccarelli, N. Cook, S. C. A. Ferse, B. Hancock, P. Harrison, M. Hein, E. Shaver, A. Smith, D. Suggett, P. J. Stewart-Sinclair, T. Vardi, and I. M. McLeod. 2020. "Coral Restoration - A Systematic Review of Current Methods, Successes, Failures and Future Directions." *PLoS One* 15(1):e0226631.

54 https://algaeplanet.com/sargablock-harvests-sargassum-for-building-bricks/.

10

THE CONFLICTS THAT CONSUME US

"Can't we all get along"? (Rodney King, 1992)

Conflicts that impact the environment can be categorized into two major categories. The first is the political conflicts over the environment—resource use, preservation versus consumption, pollution versus protection, and similar issues. These are usually peaceful, in that they do not involve an actual shooting war, but they often lead to thousands or even millions of deaths. Much of this type of conflict has been touched upon in previous chapters. The second is the human and environmental damage, deliberate or otherwise, caused by military action, actual kinetic warfare,[1] and is the subject of this chapter.

Humans are aggressive and competitive animals. For as long as we can discern, we have been practicing warfare, the organized conflict between groups. Since coming out of Africa some 70,000 years ago (or more), we modern humans (*Homo sapiens sapiens*) have spread across the planet and probably had a hand in exterminating the other hominin species (close human relatives), leaving only us.

Modern humans are no better to each other. Violent conflicts and warfare are attested in the archaeological record,[2] and all recorded history tells of frequent wars. So do the legends, stories, tales, and genealogies of all human groups except a few very small and isolated ones. Warfare seems, unfortunately, to be a regular part of human behavior. Our brains have evolved intelligence, but we still have not evolved social systems that allow us to cooperate and live together.

War is far too complicated a phenomenon to compress into one chapter, so we focus on ecological consequences, but we note that war occurs traditionally for a very few reasons, cited over and over again: Desire for more land and resources, at neighbors' expense; desire for honor and glory, particularly by the rulers of the

DOI: 10.4324/9781003560326-10

aggressor groups; desire for revenge against real or imagined damages in the past; and simple desire to fight, often by young people needing adventure, egged on by arms dealers and arms manufacturers wanting to sell and to test their wares. Whipping up hatred is always a major part of war, leading to propagandistic lying and dehumanization. This leads to a great deal of vindictive killing and destruction, over and beyond the stated goals of the war.

Wars are tragedies, not only due to the waste of human life and the environmental trauma they cause but also due to the waste of resources that could be used elsewhere for the betterment of people.

An Evolving Trauma

Prior to the development of farming, societies were relatively small. Warfare between them was small in scale and largely related to revenge for some prior action. (Some do not even call it "war," leaving that label for organized, systematic efforts.) Such warfare resulted in few casualties and little environmental trauma, though they could be devastating to the small groups involved. Still, conflict is as old as humans. There is considerable archaeological evidence of ancient warfare. Elsewhere in this book we often stress learning from the past, but in this case we can learn only from the few truly peaceable societies, such as the Semai and Temiar of Malaya.[3]

After the development of farming, human populations grew, and certain resources (e.g., water and land) became critically important. Bands of warriors, and later full-scale armies, were formed to obtain or defend them. These early armies frequently (if not always) used scorched-earth tactics. We read of armies systematically burning crops and forests and towns, cutting down orchards, destroying irrigation works and levees, butchering or looting livestock, and of course murdering and raping the unfortunate people involved—combatants and noncombatants alike. Well-managed landscapes were reduced to lifeless waste.

The Battle of Towton is a good example of some of the impacts of an early large-scale battle. This battle was fought on March 29, 1461 (Palm Sunday) and was the bloodiest battle of England's War of the Roses.[4] The records indicate that some 28,000 men died on that single day, mostly killed with swords and axes. Some of the dead were buried in mass graves spotted across the battlefield. But many were apparently just thrown into the nearby river. This would have polluted the river for a great distance downstream and would have impacted the health of the wildlife and the many people dependent on that water.

Roving armies lived off the land. The Greek general Xenophon, a careful and thoughtful writer, noted it took at least two years before an army could go through an area a second time—it took that long for the area to recover enough that the army could find enough food.[5] Of course, the local people starved after the army passed through, but that was just life. Times have not changed much in that regard. Modern war leaves devastation, even if the armies do not live off the land.

This type of warfare sometimes resulted in many casualties and produced refugees. It also caused some serious environmental damage. But the overall scale was usually small when compared to today, though such extreme conquerors as Genghis Khan, Khubilai Khan, and Tamerlane did damage, reaching levels comparable to modern wars, routinely burning cities and croplands.

Modern Warfare

Modern warfare can be defined in any number of ways. Here we think about modern warfare as it relates to environmental damage and mass casualties. The Napoleonic Wars (ca. 1800–1815) certainly resulted in large numbers of casualties, and the Russian defense against Napoleon through scorched-earth policy in 1812 resulted in substantial environmental damage. The Civil War in the US (1861–1865) is in the same general category. Sherman's march through Georgia used scorched-earth tactics whose physical effects lasted for decades.

However, it was not until World War I (1914–1918) that the scale of environmental trauma became almost mind-boggling. For four years, huge armies of millions of soldiers fought across immense areas, resulting in some 40 million casualties (dead and wounded, military and civilian).

The environmental damage done in World War I was massive. Much of it resulted from the concentrated artillery bombardments that destroyed entire regions, including cities, towns, farms, forests, wildlife, and ecosystems. Poison gas was used in some artillery shells, creating even more hazards. To this day, French farmers continue to die from World War I artillery shells, and substantial areas of northern France are off limits to people due to the ongoing danger of unexploded munitions.

World War II was even larger in scale, with even more casualties (ca. 80 million dead and wounded, military and civilian, not to mention the extermination of whole ethnic groups) and even greater environmental trauma. Large armies equipped with sophisticated weapons and aerial bombing caused massive damage. Much of western Russia was destroyed, as were large areas of China, Japan, the Philippines, Poland, Germany, Italy, and France. This damage was a global environmental catastrophe, culminating in the use of the atomic bombs on Japan in 1945.

Of the subsequent wars, the Vietnam War (ca. 1961–1975) stands out. While smaller in scale than the two World Wars, the bombing of Vietnam and adjacent Cambodia and Laos was heavier than it had been in World War II and left scars and dangers still being dealt with today. In addition, the US used a defoliant called Agent Orange to denude large areas of rainforest and mangroves, constituting an ecocide.[6] This not only impacted the plants, wildlife, and humans in the forest but has also had lasting effects on the health of millions of people (and their children) exposed, including many thousands of US service personnel. In 2012, the US began to assist the Vietnamese in cleaning up Agent Orange storage sites.

While most recent warfare is small in scale compared to the World Wars, they still have caused considerable damage, including to natural systems, human populations (including instances of genocide), and infrastructure. The current Russo-Ukrainian war (2022 [2014] –?) is a return to the large-scale destruction reminiscent of World War I. The loss of life and environmental damage is staggering.

If there was ever a full-scale nuclear war, there would be unimaginable consequences. Billions of people would be instantly killed, and massive damage would be done to property and the environment. There would also be the effects of radiation on the surviving life forms, but that would probably not matter since the explosions would send so much debris into the air it would produce a "nuclear winter" condition that would greatly inhibit agriculture and result in the starvation of most of the survivors anyway.[7]

Weaponizing Water

The use of water as a weapon of war, as a substitute for soldiers, has been all too common. These events cause damage to the environment and can kill many innocent people and animals. For example, in 1938, the Chinese destroyed a dam to stop the advancing Japanese Army, but as many as 900,000 civilians died in the flooding. The Chinese had made this mistake many times before, including a dreadfully misguided attempt to stop the Mongol hordes that actually helped the Mongols by devastating much of central China. In World War II, the Dutch breeched the levees and flooded a portion of their country to try to stop the Germans. A few years later, the Germans did the same thing in the same place to try and stop the Allies.

Also in World War II, the Soviets destroyed the great Dnieper Dam (now in Ukraine) on the Dnieper River to slow the German advance into eastern Ukraine. The resulting flood killed as many as 100,000 civilians and Red Army troops and caused considerable damage to the river environment.

Most recently, in 2023, the Russians that had invaded Ukraine destroyed the Kakhovka Dam, also on the Dnieper River, to prevent the Ukrainians from crossing it. The resulting flood killed dozens of civilians, an unknown number of Russian soldiers, and many animals. Cemeteries were also disrupted, as well as the burials of diseased cattle, raising the fear of infection. A UN Post Disaster Needs Assessment (17 October 2023) lays bare the scale of the 6 June 2023 tragedy; the assessment totaled 14 billion dollars in damage.

Danger! Do Not Enter!

In addition to the death and destruction of active warfare, there are the residual effects of left-over weapons and contaminants. Destroyed infrastructure must be rebuilt, the dead buried, ecozones restored, and debris removed. Some of the weapons used in the war remain deadly even after the conflict ends, primarily mines (land and naval), unexploded ordnance (UXO), and contaminants. Improvised Explosive Devices also present a problem. Many thousands of innocent people are killed each year by these leftover weapons.

Land mines come in a bewildering array of forms but sort into two major types: anti-tank and anti-personnel. Mines are designed to limit the movement of opposing forces in certain areas that, ideally, are mapped so the mines can later be removed. Unfortunately, such mapping is uncommon, so the location of many mine fields is unknown, especially in conflicts involving poorly trained combatants. Now, mines can be deployed remotely, making the job of finding them even more difficult.

Anti-tank mines generally require the weight of a vehicle to detonate, so they are very dangerous to farm equipment and other vehicles but not so much to people on foot. Anti-personnel mines are designed to kill individuals on foot (detonated by direct pressure, tripwires, or motion sensors) and are extremely dangerous. Among the most mined areas are Ukraine, followed by Afghanistan, Angola, Cambodia, Columbia, and Laos.

UXO typically includes artillery shells, aerial bombs, and missiles that were intended to explode on contact but for some reason did not detonate. Some are true duds that will never explode, while many others just need a bit of encouragement from disturbance to detonate. Shells containing poisonous gas also remain dangerous. People digging into the soil for some reason, such as construction or farming, can encounter UXOs, and each year dozens of farmers die from UXOs. In France and Belgium, the farmers joke about the "Iron Harvest" of UXOs they find in the fields each year. These materials are collected and destroyed at specialized facilities (they were dumped in the ocean until 1980).

Clearing land of mines and UXO is difficult and time-consuming. Both governments and NGOs (e.g., the Halo Trust) work to remove these explosives. In some areas, removal is impossible due to the vast numbers of UXO that are present, and some of the areas of the World War I Western Front in France have been placed in exclusion zones (*zone rouge*) (Figure 10.1). More recently, UXO in Vietnam has been dealt with by driving sheep and pigs through the areas to detonate mines. The unfortunate finders are then eaten. More positive is the use of giant African rats to sniff out mines; they have a special sensitivity to the scent. They are an odd but very real modern group of heroes.

In addition to the deaths caused by exploding UXOs, they also decay and release their contents of explosive or poisonous chemicals, resulting in the contamination of both soil and water. One area in France is so contaminated by arsenic from World War I UXO that vegetation still will not grow there (it is in a *zone rouge*).

Finally, contamination of various kinds results from the preparation for war, such as at weapons testing facilities, military bases and training locales, and weapons disposal areas. Of note is the contamination of large areas due to nuclear testing, such as chunks of the Nevada desert, Bikini atoll, and Semipalatinsk, the highly contaminated Russian nuclear testing area.[8]

A related issue is nuclear accidents that render areas uninhabitable. The most notable of these is the 1986 accident at the Chernobyl Nuclear Power Plant, then in the Soviet Union and now in Ukraine. Immediately after the accident and release of radioactivity, an Exclusion Zone was established with a radius of 30 km (280 km^2)

FIGURE 10.1 The extensive landscape damage at the battlefield of Verdun as it is today. Public Domain.

from the plant but later enlarged to cover some 2,600 km². Free from human influence, the area returned to being forested with considerable wildlife.[9] In February of 2022, invading Russian soldiers occupied the Chernobyl area and entrenched themselves into the contaminated soil, no doubt becoming ill from the exposure. Oops! On the subject of contamination, in 2023, invading Russian soldiers in Ukraine entrenched themselves into soil contaminated by buried infected farm animals, possibly becoming ill from the exposure. Oops again!

Long-term Impacts of Warfare

Once the dead are buried and the property damage repaired, the devastated landscape remains. Animals and plants were killed, forests leveled, habitats ruined, and ecozones destroyed—truly an ecological disaster. In theory, nature will heal these wounds in time, but sometimes human help is needed. For example, the wild European bison were exterminated by soldiers in World War I. After the war, a program to breed stock in zoos was successful, and they were reintroduced into the wild in 1929. Today, healthy populations reside in Poland and Belarus. In the late 1800s, the US virtually exterminated the American bison to deny them to Native Americans, but their population has recovered somewhat over the last 100 years.

Warfare can disrupt agricultural production and food supplies. After Rome defeated Carthage in 146 BC, the Romans are said to have salted the Carthaginian agricultural fields so nothing would grow there again (there are other examples of this). Armies will burn crops and destroy farm facilities to deny food to the civil population and enemy armies (think of Sherman's "March to the Sea" during the American Civil War). This "scorched earth" approach is often a military policy, as is reviewed by Emmanuel Kreik in a book of that title.[10] The recent disruption of Ukrainian grain exports to the rest of the world that impacted the food supply for many millions of people is another example (but now solved, at least in the short term). Again, history repeats: Russia caused a deliberate famine in Ukraine (called the Holodomor) in 1932–1933 to subjugate Ukraine by starving it into submission. Agriculture was destroyed and food sequestered. This was one of a large number of genocides by deliberate famine, chronicled by Rhoda Howard-Hassman in *State Food Crimes*.[11]

Genocide is the killing of an entire ethnic or religious group, specifically peaceful or noncombatant citizens of the killers' own country. This has a long history. War can be initiated with the goal of a genocide, and the elimination of the Jews (the Final Solution) was a major war aim of the Nazis. Other planned genocides include the Tutsi in Rwanda in 1994, the Muslims in Bosnia in the early 1990s, and the Christians in Syria in the 21st century. A genocide might also be part of a larger war, as in the case of the Armenians during World War I. In any event, any genocide is a major environmental catastrophe. The Armenian genocide, for instance, led to the abandonment and ruin of major agricultural areas that had been farmed by Armenians.

Past the trauma of war, the resulting political changes should also be considered as environmental impacts. Such changes would be less obvious than the physical damage but may set the stage for changes in policies that affect the environment.

Reducing War?

Attempts to reduce warfare and bring peace have been locally and transiently successful, but the state of the world as of 2024 shows we have basically failed. The problem often breaks down to politicians whipping up hatred to increase and consolidate their own power. Real grievances can cause war, especially civil wars that follow from a neglected and oppressed region rebelling against the national government (e.g., Myanmar). Extremist ideology—religion, communism, fascism—is the excuse for many wars. In the old days, the Vikings, Mongols, and others were more forthright about wanting loot and rape. Today, economic interests are always involved. War is not cheap, and somebody must expect a huge profit. The arms industries of the world are obviously in the running, but large-scale extractive industries also want to expand their power and reach. They easily find ambitious politicians to fund.

We cannot stress enough the degree to which extremely gentle, peaceful, well-meaning humans can be turned into crazed killing machines by political hatemongering. ENA's studies of genocide took him to Cambodia and Rwanda, where some of the most famously peaceable people in the world committed some of the worst genocides. It is important to note that war and genocide have taken place under fascist, communist, religious, capitalist, and mixed regimes. There is no simple solution. The one common theme is that in every case, extreme ideological rhetoric took over, and sober voices were silenced. We do not fully know what to do about this.[12]

Cures must involve better world governance, from treaty-making to international policing, but in the end we must deal with the ease of whipping up hatred in a sadly contentious species. Humans can break either for peace or for war, just as they can break either for violence and hate or for decency and civility in everyday life. Teaching people to get along, and also teaching the full extent of the horrors of war, is a necessary start.

Notes

1 Crawford, Meta C. 2022. *The Pentagon, Climate Change, and War*. Cambridge: MIT Press. She argues that the damage done to the environment, including climate change, one of the major issues.
2 Arkush, Elizabeth N., and Mark W. Allen. 2006. *The Archaeology of Warfare*. Gainesville: University Press of Florida.
3 Dentan, Robert Knox. 2008. *Overwhelming Terror: Love, Fear, Peace, and Violence among Semai of Malaysia*. Lanham, MD: Rowman and Littlefield.
4 Goodwin, George. 2012. *Fatal Colours: Towton 1461-England's Most Brutal Battle*. New York: W. W. Norton & Company.
5 Xenophon. 1998. *Anabasis*. Cambridge, MA: Harvard University Press, Loeb Classical Library.
6 Zierler, David. 2011. *The Invention of Ecocide: Agent Orange, Vietnam, and the Scientists Who Changed the Way We Think about the Environment*. Athens: University of Georgia Press.
7 Coupe, Joshua, Charles G. Bardeen, Alan Robock, and Owen B. Toon. 2019. Nuclear Winter Responses to Nuclear War Between the United States and Russia in the Whole Atmosphere Community Climate Model Version 4 and the Goddard Institute for Space Studies Model E. *Journal of Geophysical Research: Atmospheres* 124(15):8522–8543.
8 Yan, Wudan. 2019. "The Nuclear Sins of the Soviet Union Live on in Kazakhstan." *Nature* 568(7750):22–24.
9 Oliphant, Roland. 2016. "30 Years after Chernobyl Disaster, Wildlife Is Flourishing in Radioactive Wasteland." *The Daily Telegraph*, April 27, 2016.
10 Kreike, Emmanuel. 2021. *Scorched Earth: Environmental Warfare as a Crime against Humanity and Nature*. Princeton, NJ: Princeton University Press.
11 Howard-Hassmann, Rhoda E. 2016. *State Food Crimes*. Cambridge: Cambridge University Press.
12 See Anderson, E. N., and Barbara Anderson. 2022. *Sustaining Social Conflict*. Lanham, MD: Rowman & Littlefield. Genocide expert Erwin Staub has worked hard on finding ways to make genocide less likely, and come up with many good ideas: Staub, Ervin. 2011. *Overcoming Evil: Genocide, Violent Conflict, and Terrorism*. New York: Oxford University Press.

11

WE CONTROL OUR FUTURE

"And the Men [and Women] *Who Hold High Places Must be the Ones that Start"* (lyric from "Closer to the Heart" by Rush)

We have made quite a mess of things. The forests are disappearing, a mass extinction is underway, the oceans are dying, the air and water are poisoned, and millions of people are starving. Despite the many warnings and dire predictions, we continue to trash the planet and, in the process, have made our ability to live on it more difficult.

Most of the naysayers naysaid due to their desire to make money. Energy companies, polluters, big business—all work on that shortsighted goal. This is made possible by politicians being in the pockets of those businesses, preventing any real action on the climate that might cut into profits. In 2006, Al Gore warned us in his film documentary and subsequent book, *An Inconvenient Truth*. Even in 2024, Republican candidates have openly declared climate change a "hoax." This is totally irresponsible and has deadly consequences. Deaths from extreme heat are now in the hundreds of thousands worldwide, and global climate change would surely have been mitigated long ago if it were not for political disinformation.

So, What Can Be Done?

Houston, we have a problem.

(paraphrased from the statement made by astronaut
John Swigert aboard Apollo 13, 1970)

DOI: 10.4324/9781003560326-11

Industrial society will have to change. We can no longer just maximize throughput, producing more and more and then throwing more and more of it away. We will have to use less and use it *much* more efficiently.

Traditional societies thrive with less and have good lives. We cannot go back to old-time agrarian societies, but we will have to build on their approach, using less to do more, inventing workarounds, and above all, conserving everything possible for use. Using the most that was possible, in order to get maximal efficiency and have a reason to save everything, is probably the most important single thing we can learn from them.

In fact, we have the technology to do this, and there are many practical ideas. We lack only the political will and the economic structure that would allow us to evaluate costs and benefits in ecological terms rather than strictly in terms of how much stuff is used (and wasted).

There are a number of things we can do to deal with these problems, if we have the will. It is very clear from the recent fires in Maui and hurricane Hillary in Florida that we are currently unprepared to deal with the consequences of the climate crisis. But the cost of future wildfires, severe storms, droughts, and sea level rise will far and away exceed the cost of dealing with the issues now. Informing and educating people to understand the basic issues and the urgency of action is among the first things to do.

The Information Battle

Unfortunately, much of the right-wing media in the US has convinced their viewers and listeners that climate change is not real and that the liberal left is lying about it. The left claims just the opposite. What can we believe? We have already discussed the lies told by the companies doing the emissions and polluting. They deny the science because they and the politicians they control do not want to disembark their gravy train.

The science deniers parade "experts" that manipulate the inherent uncertainties of science (see Chapter 1) and twist it into a false argument that the science is wrong. They misuse statistics to mislead the public about the real problem. Unfortunately, there is not a corresponding response by public relations firms representing science.

Fighting the lies can be done only in the context of fighting all the corporate and political lies that now dominate the media. Racism, anti-vaxx, anti-public health, the "tobacco is safe" and "all chemicals are safe" lobbies, and the other ongoing lie machines must be called out, fully investigated by legislatures everywhere, and shut down as much as possible by anti-fraud and anti-false-advertising laws. These are *not* matters of debate. The ones that are direct lies by corporations about their products are *not* protected speech. Lies and bribery to politicians are *not* protected either. Different nations have different laws, but all have laws against at least some open corruption. The giant fossil fuel corporations and their allies—especially big agribusiness—have corrupted whole governments around the world.

Also, it is necessary for citizens to do the research on their own. They must familiarize themselves with the issues and facts. Armed with this information, people can make informed choices about what actions to take. A major work on how to fight disinformation and misinformation is *Foolproof* by Sander van der Linden.[1] This is a book that should be widely read and used in teaching.

Is It Possible to Reverse Climate Change?

The emission of GHGs into the air is the primary cause of climate change. The climate will continue to get hotter until we stabilize atmospheric carbon, and then we might have a new "normal" climate that will be hotter than today. To reverse the effect and return the climate to an earlier time (say, 30 years ago) would require the removal of massive amounts of carbon from the air, equal to the emissions from the last 30 years plus the current yearly emissions. Although this seems to be a bridge too far, there is some movement in that direction.

Natural Carbon Capture

Perhaps the best way to remove carbon from the air is by planting trees and other vegetation that will absorb and store it in their tissues. Reforesting natural forests is the greatest hope and would remove 25 percent of the aerial carbon,[2] but much of the world's former forest land is now committed to agriculture. Tree plantations can be useful, but they store less carbon and are usually cut young, liberating the carbon again. Natural forests also pack much more storage into a given area because they have a wide variety of species with different requirements. Also, regrowing forests takes a while to start net absorption of carbon; for a few years, decay of old wood and litter outweighs the advantage.[3]

There are many neat tricks for blotting up aerial carbon. An even 100 are included in the wonderful 2017 book, *Drawdown*, edited by Paul Hawken.[4] Hawken picked out many of the best ways to use vegetation to absorb carbon, including mangroves and other coastal wetlands; agricultural savannahs, with fruit trees and other useful trees; bamboo; hemp; and perennial grasslands (plus various ways of making cars, ships, buildings, and other carbon sources more efficient). Many of the techniques involve restoring traditional agriculture and other traditional landscapes. Some involve sea and wetland development since mangroves, seagrasses, coastal marshes, and the like are highly productive and dynamic environments. Preserving such coastal wetlands could trap as much as two percent of the GHGs in what is known as "blue carbon."[5] In 2024, CNN produced a documentary on Blue Carbon.

Another set of useful ideas is described by Richard Conniff.[6] These include crushing rock to expose more surface area to the air since the weathering of rock is second only to absorption by plants as the natural way to pull carbon from the air. However, this would take too much energy to be practical unless a better method of crushing it is developed.

Still another set of ideas, focused on energy supply and transfer, is covered by Mark Jacobson.[7] These appeal to technology that has been or could be developed but is not yet available for ordinary use, with the hope that it will be made widely available soon.

Synthetic Carbon Capture

Another possible way to remove carbon from the air is synthetic capture and storage. Some carbon can be captured during manufacturing processes. Other carbon can be captured directly as they are emitted, such as using carbon scrubbers or absorbers on smokestacks. While this is perfectly possible, it is rarely done.

If CO_2 is already in the air, there are technologies to capture and remove the gas, but in most cases, they are currently too expensive to be very useful. However, a new method for removing CO_2 from seawater has been developed[8] and appears to be more efficient than existing systems for removing it from the air. By removing carbon from seawater, it would allow the oceans to absorb other carbon.

Captured CO_2 is typically stored in underground geological formations, though leakage is an issue. Carbon might be removed from seawater by ships and converted into a fuel used by the ship.

Restructuring Energy

For most of human history, people used energy derived from burning plant materials and using animals for labor. For example, in 1850, about 91 percent of the energy used in the US came from burning wood and other biological materials. By about 2000, most of the energy used by humans originated from fossil fuels, such as gas, coal, and oil, and in the US, fossil fuels provided some 81 percent of the energy used. Given that fossil fuels are the primary contributor to GHGs and drivers of climate change, there are efforts to develop alternative energy sources to replace fossil fuels. Some fear that this effort will fail and predict the demise of industrialized societies as a result.

We use energy in two major forms: electricity and liquid fossil fuels to power cars, trucks, planes, and ships. Most of the electricity we use is produced in power plants fueled by coal, which pollutes the air and land (soot, fly ash, and nasty chemicals) and is a major emitter of GHGs. In the US, coal is being replaced by natural gas, less polluting but still a carbon emitter. But coal-fired power plants are still being built elsewhere, especially in China and India. Still, most new power plants in the world are using green renewables.

It is clear that we need to very rapidly move away from fossil fuels. However, since fossil fuel companies get huge government subsidies, they are actively resisting the transition to green renewables. If politicians stood up to stop the subsidies, the energy market would naturally move to cheaper green renewables. The Catholic Church has taken a lead. Pope Francis came out with his own strongly worded encyclical,[9] and the church sponsored a publication (not all by Catholics) summarizing the problem.[10]

On the plus side, the International Energy Agency (IEA) predicts that global demand for fossil fuels will peak in 2030.[11] Still, fossil fuels and the products made from oil and coal remain necessary for a variety of purposes. The problem is over-production, overuse, and overdependence. It seems likely that overdependence and oversubsidizing of any resource would be equally pernicious; the problem is the addiction, not the drug. Interestingly, there is a new approach to re-purpose coal-fired power plants, using their steam generators and other infrastructure but with large batteries charged by solar or wind to produce power at night. Cool!

The future seems to be all-electric, or at least mostly electric. Technologies to power transportation using electricity can be seen everywhere. Electric vehicles or fossil fuel/electric hybrids, including cars and trucks, are now common. The problem is that much of the electricity used in these vehicles is generated by fossil fuel power plants, so they are not really "green." But this will improve as we develop the capac-ity to generate sufficient electricity from renewable sources. We are on the right track.

Another way to increase available power is to use less of what is generated and so reduce the demand for new power plants. This can, and is, being achieved by increasing the efficiency of existing technologies: higher gas mileage, better refrig-erators and air conditioners, more efficient lightbulbs, adjusting thermostats (many restaurants and hotels are freezing inside), and more vegetation in cities. Energy expert Vaclav Smil said in an interview in 2022:

> In managing our energy affairs we should constantly favor doable steps: not wasting 40 percent of our food grown with high energy expense, not to heat or cool the universe in poorly designed but oversize houses, not to waste fuel and materials driving SUVs (nearly two tons of mass to move, usually, a single per-son), not to design cities that demand lengthy commutes, not to keep amassing rarely used products, not to travel mindlessly.[12]

There are a number of clean and green, renewable, and less expensive (once sub-sidies are removed) alternatives to fossil fuels, some of which have been used for centuries. While the transition from fossil fuels will result in a loss of jobs in those industries, those workers could be retrained into new clean energy jobs. Such tech-nological shifts have occurred many times before. For example, when the US (and other countries) shifted from a horse-based transportation system to one of auto-mobiles, blacksmiths and stable owners came under pressure, but they adapted by shifting to auto repair work in garages. Clean energy is at our fingertips.

Dam Power

Hydroelectric power is a long-established method to generate electricity, currently accounting for about 6.3 percent of the electricity used in the US. Hydroelectric power is clean and renewable, but not without its issues. The construction of a dam creates a reservoir that floods natural ecosystems, impacts fish populations, and reduces biodiversity. It also requires that people living in the footprint of the reservoir move to new homes (in the 1930s, the TVA displaced some 125,000 people for their reservoirs). On the Great Plains, dam and reservoir construction disproportionately affected Native American people living in that region. Archaeological sites are also inundated and damaged or destroyed. In addition, reservoirs emit methane, and water is lost through evaporation. Finally, reservoirs will eventually silt in and negate any power production, and all have the possibility of collapse (that would be very bad!). We have reviewed the problems with dams in Chapter 6.

Sunny Side Up

Solar is a category of technologies to convert the energy from the sun into electricity and currently provides about three percent of the electricity used in the US (this percentage is increasing). Some solar power plants use mirrors to focus sunlight onto a spot where it heats water that makes steam to turn turbines to create power. However, most solar projects use photovoltaic panels to directly convert sunlight into electricity. These come in sizes ranging from gigantic ones covering thousands of acres to the small panels on the roofs of private homes. Photovoltaic panels are rather inefficient in conversion, but that too is rapidly improving.

The biggest problem with solar is inconsistent production. Solar cannot produce power at night or in bad weather, an issue mitigated by storing the power produced in the day in large-scale batteries for use at night and other times. Such battery storage is already being used, and the efficacy of the batteries is improving, making solar more attractive.

Solar panels can be placed almost anywhere, including every roof, parking lot, and anywhere else needing shade. Solar panels could also be placed over canals and reservoirs—they would greatly reduce evaporation losses by shading the water.

However, solar is not without its issues. Solar farms require a great deal of land that is damaged by their construction, maintenance, and infrastructure. Many solar projects are in deserts, areas that have considerable sunshine and seem to be considered expendable (nothing lives in the desert, right?). Thus, considerable damage is done to desert ecosystems by solar projects. In addition, if birds fly too close to the solar plants using mirrors, they can be incinerated.

Another issue is the disposal of solar panels once they reach the end of their use-life. What happens to the toxic chemicals used to make the panels? Can they be recycled, or will they be put in landfills with their contents bleeding into the soils and water tables? We do not yet know.

Blowin' in the Wind

Wind energy currently produces about nine percent of the electricity produced in the US. Massive wind farms have been constructed in windy areas, and gigantic wind turbines can be seen slowly turning their built-in generators. In Europe, large wind farms have been built in the North Sea. Smaller wind generators can be used for businesses and homes. Using wind power is not new; many of the pumps in wells across the US were powered by windmills for hundreds of years (Figure 11.1). The Dutch have used wind power for many centuries. Tasmania generates all of its electricity through wind energy.

Wind power has several major drawbacks. First, it is unreliable since if it is not windy, no power can be generated. Like solar, this issue can be mitigated by storing power produced during windy periods in large-scale batteries for use at other times. The other major issue is the damage the construction and operation of wind farms do to the environment. Like solar, onshore wind farms require a great deal of land

FIGURE 11.1 American Style Windmill. Photo by Agnes Monkelbaan (Wikimedia Commons, Public Domain).

that is modified for the towers, with roads between them, power cables connecting them, and transmission lines to connect to the grid. In addition, wind turbines kill many birds and bats, but this has been solved, in concept, by better design and placement.

An unusual upside to both solar and wind facilities is the ground beneath the panels and turbines. In some cases, the land is suitable as pasture, and grazing animals can be raised there to produce meat and reduce weeds—a method called "Solar grazing." Leasing the land to a rancher can also generate some income for the solar or wind facility.

The Tiny Mighty Atom

There are two basic types of nuclear power generation: fission and fusion. Nuclear fission is the only one currently viable and is widely used but has some issues. Fission involves splitting atoms (mostly commonly uranium), which generates heat that can be used to run steam generators. However, if not tightly controlled, the heat from fission can increase to the point of melting the containment building and releasing radiation into the air and water, such as happened at Chernobyl and Fukushima. This fear of nuclear accidents has made such power plants undesirable. In addition, the fuel rods used in fission plants eventually lose their efficiency and must be replaced, and dealing with the spent rods is an issue. Moreover, nuclear fission power plants are expensive to build and still have a substantial carbon footprint (just think about all the methane-emitting concrete), though far less than any fossil fuel plant.

An innovation in fission power plants is the development of a new type of fission reactor using thorium mixed with salt as a fuel (molten salt reactor, or MSR). MSRs promise to be much safer, and due to their low-pressure operation and safety features, they cannot melt down, explode, or release radiation into the air or water. Seemingly much better.

Also under development are microreactors, small nuclear power stations designed to provide electricity to a military base or small town. Such reactors have long existed, used by the Navy in submarines and planned to be used by NASA in space, but now may be used for civilian purposes. Such reactors could be built in factories and placed most anywhere they are needed.

Nuclear fusion takes a different approach, fusing atoms into heavier elements and thus creating heat to run steam generators. In theory, fusion plants will be less hazardous and cleaner but will have a big carbon footprint. However, fusion power is still in the developmental stage, and it is not known when such plants will come online, perhaps sometime around 2050.

Other Clean and Green Sources

There are several other sources of clean and green renewable electrical generation. Geothermal power plants tap the heat of the earth in volcanically active areas to drive steam generators and currently provide about 0.4 percent of the electricity used in the US Geothermal plants do not use fossil fuels, and their construction and maintenance have only a small carbon footprint, but they can only be built in certain places.

A new avenue of electrical production is the harvesting of energy stored in the ocean. This is a similar concept to wind power but using ocean waves, currents, and tides to run generators. Power can also be derived by exploiting differences in temperature and salinity. This requires that the facilities be anchored to the sea bottom and that cables connect them to the grid, each of which would cause some impact to marine life. Some such facilities are now operating, and the potential seems promising.

There are also several green renewable fuels for vehicles. One is hydrogen used in fuel cells; it is very clean and leaves only water as "exhaust." The downside of hydrogen is that, today, it is mostly produced from natural gas (called gray hydrogen) or by capture from other processes (blue hydrogen). Hydrogen produced by electrolysis is called green hydrogen but requires electricity, currently produced mostly by fossil-fueled power plants. The production and use of green hydrogen is the goal but will require renewable energy to be useful as a fuel. In addition, the distribution infrastructure for hydrogen is so far very limited.

Biofuels are another green renewable fuel and come in two major forms, bioethanol and biodiesel. Bioethanol is typically produced from processing corn or other biological material and can be used as is or added to gasoline. Biodiesel is produced from fats and oils and can be used alone or added to diesel fuel. Biofuels theoretically reduce the consumption of fossil fuels, but their production has a carbon footprint. Another major issue is the use of biomass (e.g., corn) that could be used to directly feed people. To avoid this issue, "second-generation" biofuels are made from agricultural waste rather than products humans can eat.

Another possible fuel source is a new technology that can extract carbon from seawater and convert it into fuel. This process is not yet perfected, and research is ongoing. This could revolutionize fueling ships at sea.[13]

Adapting to Climate Change

It appears unlikely that we will be able to engineer a return to a cooler Earth. Even the hope of limiting warming to less than 2°C seems almost unachievable. While we need to keep the situation from getting worse, we also must develop strategies and methods to adapt to the new climate regime we are making. Some believe that it is certain that our contemporary industrial society will disappear within the present century. The goal, then, is to replace it with a new society based on efficiency and sustainability rather than on maximizing throughput at all costs.

To keep things from getting worse, we must confront, directly, the world fossil fuel industry. Oil now runs whole nations, mostly small producer ones, but also some quite large states. Oil dominates politics in democracies (the US), fascist dictatorships (Russia), theocracies (Iran), kingdoms (Brunei, Saudi Arabia), socialist dictatorships (Venezuela), and indeed states with every imaginable form of government. The opposition of "communism" and "capitalism" is irrelevant in the contemporary world; the opposition of the fossil fuel industry and the survival of life on Earth is the great contradiction.

Giant corporations can corrupt whole nations. An excellent backstory on these matters is *The Plundered Planet* by Paul Collier. Another with major remedies recommended is *The Value of Everything* by Mariana Mazzucato, who has been a major contributor to international debates on how to square development with equity. James Wood's great work, *The Biodemography of Subsistence Farming* remains basic for all studies of environmental decision-making, including conflict.[14]

The political dominance of oil (and coal in some places) must be eliminated. This means removing subsidies and, instead, forcing the corporations to pay for the damage done. So far, they do not even pay their fair share of taxes and are notorious tax evaders in the US. They apparently do not pay much anywhere else either, except possibly Norway.

In a fully marketized economy, competition would force firms to be more efficient. They would have to pay for environmental damage, and they would be rewarded for making maximal best use of resources.[15] Unfortunately, the world has converged on an economy that is neither free market nor socialist: it allows giant firms (whether independent or governmental) to produce and use resources, but then uses tax money to do the cleanup. This is why accounts of China's "communist" economy sound so strikingly familiar to Americans used to the "free market" in the US. The fact is that both countries, and most of the rest, have converged on the idea of subsidized production and taxpayer-funded cleanup.

Only phasing out the subsidies and forcing the firms to clean up their own mess will solve the problem. This, however, requires that we have alternatives on hand. Unfortunately, the world now depends on cheap oil and gas. Imagine the effects of pricing gasoline at its actual environmental cost, quite possibly hundreds of dollars per gallon. The disturbing fact is that sooner or later we *will* pay that. We are kicking the can down the road. Our grandchildren will pay as they try to survive climate change, polluted waters and lands, poisoned air, and governments ruined by subsidy and cleanup costs.

In the short run, then, we are stuck with government remediation. Tax money will take care of what can be done, while we move as fast as possible toward efficiency in using materials, planting trees and grasses, using less disposable stuff, and making firms pay the costs of the benefits they get from the public.

Whispering Echoes

State societies long before ours were able to cope successfully with climate change, and the whispering echoes of their stories can inform us. Long-running systems invariably encountered climate changes, but even rather short-lived states sometimes faced sudden climate stresses and survived. Hoyer and colleagues[16] discuss the success of the Byzantine Empire in surviving the changing climate that coincided with the decline of the Roman Empire and then the major cooling and drying event of AD 536 to 538, when volcanoes injected enough dust into the world atmosphere to cool the planet.

On the other hand, China and many other states and regions were thrown into chaos by that change. Hoyer and colleagues note the long survival of the Ottoman Empire, to which one can add the incredible tenacity of the Choson (Yi) Dynasty of Korea, surviving 600 years despite the Little Ice Age. The Little Ice Age, for instance, led to sharp cooling periods between AD 1300 and 1800 that devastated whole countries, from China to Europe.[17]

The Ming Dynasty in China survived several major climate shocks, but finally a mix of dynastic decline and intensifying Little Ice Age cold doomed it. Climate troubles had much to do with the fall of other Chinese dynasties, but they were affected fatally only when they were weakened by rebellion and corruption.

The ancient Maya, for all their successes in farming, art, astronomy, and the like, ultimately failed to adapt to drought. Their population contracted, and their complex society evolved into a much simpler one. The Maya adapted by abandoning their cities and complexities and becoming small-scale subsistence farmers. We may find ourselves on a similar trajectory.

Those who say we can adjust to what is already an extreme change should consider those works. We are in much worse shape than those nations were then because we are already so overstretched in regard to water, farmland, and other resources.

Thus, complete chaos is also possible. If the great issues of climate change, water shortage, farmland shortage, and decline of resources are not addressed, the world will slip after 2050 into rapid economic decline similar to the 1930s depression. This will almost certainly lead, as the 1930s led, to world war. Unlike World War II, which could be kept within some limits, a World War III could exterminate the human species and most other life forms.

Most likely is a middle ground. Small, agile nations with educated citizens that trust in science will survive and manage. Large nations that fall under authoritarian regimes will slide into breakdown and violence.

Climate Action and Mitigation

Despite the constant noise of the deniers, there is a clear consensus that climate change is a threat, a crisis, and that action is urgently needed. The issue has been taken up by a variety of organizations, from the UN to local governments and even individuals. As a result, a number of international accords and treaties have been made, but enforcement is difficult. China continues to be the largest emitter of GHGs, and they continue to build coal-fired power plants despite their treaty obligations. Still, there is some progress, as noted below.

International Accords on Climate

In 1987, many countries agreed to the Montreal Protocol to phase out ozone-destroying chemicals such as chlorofluorocarbons used in refrigeration units. This treaty was quite successful and reduced GHG emissions.

In 1992, the United Nations Framework Convention on Climate Change was formed, and its resultant treaty (signed by 154 countries) called for scientific research, yearly meetings (called the Conference of Parties [COP]), and future agreements to lessen climate change and insure sustainability. The annual UN Climate Change COP meetings began in 1995 (COP1 in Berlin) with the goal of dealing with climate change. At the COP3 meeting, the 1997 Kyoto Accords were agreed to and set a country-by-country limit on GHG emissions. The US did not join this treaty, and its success has been limited.

The 2011–2015 COP meetings resulted in the 2016 Paris Climate Accord (COP21), an agreement to limit GHG emissions to keep the rise in global temperature no more than 1.5°C. Unfortunately, none of the 194 signatories are meeting their goals, and some (e.g., China) are increasing their GHG emissions. Under Donald Trump, the US left the accord in 2017, but the US under Joe Biden rejoined it in 2021.

At the COP26 meeting (2021), the Glasgow Climate Pact was negotiated and agreed to by its 197 attending parties. The result was an agreement to reduce the use of coal, although India and China weakened the final version. At the COP27 meeting (2022), it was agreed that poor countries would be compensated by rich countries for climate change damage. The COP28 meeting (2023) resulted in an agreement by some 200 countries to reduce dependence on fossil fuels and to triple the use of renewable energy sources by 2030. It is a start.

In 2009, the Maldives, an island nation under threat from sea level rise, formed the Climate Vulnerable Forum (CVF), a partnership of countries disproportionately impacted by climate change. The goal of the CVF is to address the impacts of climate change on susceptible (e.g., poor and low-lying) nations and pressure the rich nations to accept responsibility for the consequences of climate change.

What Can Individual Countries Do About Climate?

The major problem in getting nations and governments to do anything about the climate crisis is that most of those governments are complicit in the problem to begin with. The corporations and politicians are making too much money to want to change things. In some cases, however, public pressure has had some effect, and there is some hope.

First, countries should adhere to their climate treaty obligations, cut emissions, and move to clean and green energy. Countries should also require their industries to clean up their mess and maintain their facilities. In the realm of the other issues, countries can make laws, or at least enforce existing ones, on pollution. They could give incentives (e.g., tax credits) to green projects. They could expand their parks and other protected areas. They could prevent the overexploitation of resources such as timber and water. There could be investment in efficient public transportation. Countries can also develop programs of landscape restoration and reforestation. For example, Germany is requiring the conversion of some farmland back into forest. So much could be done if there was the will.

In the largest investment to date, the US passed the Inflation Reduction Act in 2022, which provided hundreds of billions of dollars for the development of green energy and other climate initiatives. In addition, US emissions are declining.

In the end, though, the United Nations or a stronger international unity must step up to the plate. Jonathan Blake and Nils Gilman argue, in *Children of a Modest Star* (2024), that we must have a world government that at least has the power to regulate the most extreme environmental damages, from climate change to ocean management to deforestation.[18] These are problems that are worldwide in scope and not controllable by any one country. Blake and Gilman do not argue for a world state, only for a technocratic regulatory agency that can keep humanity alive.

What Can Local Governments and NGOs Do about Climate?

Action by local governments (e.g., states, counties, and cities) can make a difference by reducing their carbon footprint. Government vehicle fleets can be converted to green fuels. Parks can be created or expanded. Trees planted. Better insulation could be required on new buildings. Perhaps some funding could be provided to building owners and homeowners to lower their energy requirements. The list goes on and on.

There are also many environmentally focused NGOs that do great work. Conservancy organizations buy land to create preserves or to expand existing parks or wilderness areas. Other organizations exist to rehabilitate (or "rewild") damaged lands. Local governments can do the same sort of restoration work on their properties (recall the Los Angeles River project from Chapter 3). All of this reflects the growing public support for action.

What Can Individual People Do about Climate Change?

There are many small changes that individuals can make to help reduce their carbon footprint that collectively can make a significant impact on GHG emissions (and lower your monthly expenses). Among the many possible actions, you can reduce the amount of electricity you use (so it is available for green uses) by using more efficient lightbulbs, lowering your thermostat, using less hot water, and turning off electric devices you are not using. You can use less fossil fuel, drive your car less (carpool, use public transit, bike, or walk), and/or live closer to work or work from home more. Perhaps install home solar or wind equipment.

Plant shade and fruit trees to shade your house and to provide food that does not require industrialized farming to produce. Plant a garden in the yard. Reduce the amount of meat you eat, especially beef. Recycle what you can; despite its inefficiency, it still helps.

If you own a larger property that has degraded, develop a program to return it to its original wild condition. Reestablish streams, grasslands, and forests. Set aside property as nature reserves (recall Morgan Freeman's bee sanctuary from Chapter 7).

All of this is doable and useful but the most impactful thing an individual can do is to pressure their politicians to put the planet over their lobbyists and to support climate legislation and initiatives. Little will happen unless there is political will.

Even a single person can make a huge difference. For example, Greta Thunberg of Sweden became a global climate icon at age 15 when she protested Sweden's lack of climate action in front of the Swedish Parliament (Figure 11.2). She rapidly

FIGURE 11.2   Greta Thunberg outside the Swedish Parliament building, 2018 (sign reads "School Strike for Climate). Photo by Anders Hellberg (Wikimedia Commons, Public Domain).

became a world-renowned activist, addressing the UN Climate Change Conference in Poland in 2018 and the United Nations Climate Action Summit at the UN General Assembly in New York in 2019. In her UN speech, Thunberg chastised member nations for their lack of action on climate issues, exclaiming, "How dare you?" a statement that made the news worldwide and raised awareness of the issue. One single teenager did this on her own.

Another example is Bill McKibben, a leading environmentalist who in 2023 founded the "Third Act" organization to inform and involve older Americans around the climate crisis. McKibben also wrote one of the first books, "The End of Nature" (1989, with new editions), on climate change intended for a general audience.[19]

Climate Justice

Most of the people impacted by climate change had little to do with the creation of the problem. They reside in poor and underdeveloped countries and economies that do not emit many GHGs (the ten largest economies in the world emit 60 percent of the GHGs). Having little responsibility for the problem, these people face many of the most severe consequences and costs. Who is responsible for paying the cost of climate change mitigation and adaption?

Say you have a big party at your house and have a bonfire in your backyard. Somehow, the bonfire catches your house on fire, and that fire spreads to your neighbor's house. Who is responsible for paying for the damage to your neighbor's house? You are! Hopefully you have insurance. If the industrialized societies caused the bulk of the climate problems, who should pay for the impact on other countries? Simple answer, but complicated politics.

At the COP27 meeting (2022), it was agreed that poor countries would be compensated by rich countries for climate change damage, and the US contributed one billion dollars to help fund it (not much but a start). In 2023, the US established the White House Office of Environmental Justice to deal with these issues.

One of the components of climate justice is the rights of Indigenous peoples whose lands have been taken and/or destroyed. Mitigation for these wrongs includes the establishment of land set asides (e.g., reservations, preserves, and the like) both to protect the Indigenous people and their lands from exploitation by outside interests.

Another component is to reduce and regulate projects that do environmental harm; in essence, to reduce "development." However, while this would benefit the environment, many countries do not see it as justice and want their "piece of the pie." In these cases, developing countries would be deprived of much-needed industry, jobs, and funding, thus keeping them underdeveloped and poor while the rich countries continue to prosper. The rich countries reap the benefits from trashing the planet, while the poor countries are prevented from doing so. Does not seem fair.

But It's Not Just Climate!

Western society has a "throw away" worldview, one that overconsumes and then discards. This is good for industries that produce low-quality products designed to quickly break so another will have to be purchased. But this is very bad for the planet, especially since all that stuff ends up as trash, mostly in landfills but also scattered about everywhere.

As resources dwindle, we will likely begin to fight over them. Water, land, oil, and metals (think of the current competition for lithium) will all be targets. Warfare will lead to massive damage to both the natural and cultural environments and likely to genocides, some of which are ongoing today.

All of these problems, plus the climate crisis (see above) and inequality (see below), are subjects of the massive lying and misinformation campaigns led by the very people (recall that the US Supreme Court held that corporations are "people") behind the problems in the first place. Climate change is a hoax! Tobacco is good for you! Consume more! Science is bad! Anyone with a functional brain would know these are false claims.

There is, thus, a major need for reaffirming science and making actual scientific results widely available. David Lazarus points out that since the giant firms that spin these claims are covered by fraud and false advertising laws, their anti-science lies could be banned without damage to "free speech" as normally interpreted.[20] He argues for extending those rules to politicians, at least when they are talking about corporate interests.

All Things Being (Un)Equal

A huge issue is inequality, which, at its current level in most nations worldwide, is probably not sustainable without some sort of collapse or revolution.[21] Inequality, as usually understood, means income inequality; the one percent having more than the remaining 99 percent. Inequality in access to resources, to recreational opportunities, to education, and to other public goods is also important, needing more study. This disparity is often the direct result of heavy subsidies for addictive products—from tobacco and alcohol to oil, coal, palm oil, and other commodities that have turned deadly. Again, the problem is not the commodities. Tobacco should probably be banned outright everywhere, but alcohol, opiates, oil, coal, sugar, and other commodities in need of control have their uses. The problem is our addiction and our feeding it by enormous taxpayer subsidies, giveaways, sweetheart deals, and special favors to the suppliers. This echoes through the economy, creating inordinate wealth for the banks that finance them and the manufacturers that make them. It even has something to do with the wealth of movie and sports corporations—which, along with dangerous drugs, are the suppliers of desperately needed escapes.[22]

Industrialized societies often use the "Plantation Model," in which the rich "owners" exploit (and sometimes literally enslave) the poor "labor" to gain and maintain power and wealth through the control of large businesses. A "Social Democracy Model" is one in which power and wealth are dispersed, such as in small businesses (e.g., family farms, small companies). The latter model has less inequality but is uncommon.

Current levels of inequality, throughout history, have always led to massive corruption and violence. These are all too visible today. So are disease, poverty, lack of access to water and food, and other injustices, as memorably laid out by Richard Wilkinson and Kate Pickett in *The Spirit Level*. John Rawls has called through many decades "justice as fairness," seeing treating people fairly and according to the Golden Rule as the basis of social morality.[23] Far too much intellect is trapped in poverty.

Inequality leads to injustice, and that leads to massive rebellions and wars and the collapse of the inequal systems.[24] This is part of wider resilience cycles; all animals, and even plants and bacteria, go through cycles in which density-dependent problems build up until the populations decline. They then usually rebound.[25] This may happen, but humans may die out instead.

Justice for all humans and at least some consideration for other species is absolutely necessary, as shown by genocide studies as well as the ecological literature; a masterful review by Rachel Killean and Lauren Dempster[26] cited hundreds of references, covering a large scholarly literature, to the problems of war, mass murder, conflict in general, and extreme inequality and unfairness at all levels and involving potentially all species.

The current governments of the world are totally failing in these areas. The problem was captured 2,000 years ago by Aesop's fable of the mice and the cat. The mice want to stop the cat from eating them. One comes up with the brilliant idea of putting a bell around the cat's neck, so the mice can hear it coming. The leader of the mice allows that this is a splendid idea, then asks, Okay, now, who will bell the cat? Silence. Today, the cat—the world cartel of fossil fuel companies—makes mincemeat of politicians who try to bell it. No current politician in major countries seems willing to try, despite success in a few minor cases (Scotland and Norway).

What Can Governments Do?

The question is not what countries *can* do; it is what they are willing to do. There are many things that can be done, but as with dealing with climate change, the major problem is that most of those nations and governments are complicit in the problems to begin with. In some cases, however, public pressure has had some effect, and there is some hope.

For example, countries can do a much better job of regulating the exploitation of forests and fisheries. They can require more extensive recycling and enforce laws on the disposal of trash. They can bell the polluters. But these steps require political courage, a commodity sorely lacking in many governments.

How Can Individual People Help?

There are many things that individuals can do to address the various non-climate issues we face. Probably the most important is to pressure your government to act responsibly. A few other steps are noted below but there are many more.

Use less and be more self-sufficient. Reduce water consumption by having climate-appropriate landscaping. Plant trees for shade and food. Have home gardens for fresh vegetables, like the victory gardens of World War II. Eat locally and buy produce from your farmer's market (this lowers transportation costs and supports small farms). Reduce the amount of meat you eat, especially beef. Eat what you buy (huge quantities of food, perhaps 40 percent, are wasted each day, and discarded food is a major source of methane in landfills). Millions of edible pumpkins end up in landfills, so make your pumpkin into pie at Halloween. Use fewer plastic products, avoid single-use bags, water in single-use plastic bottles, and plastic pod detergents (the plastics dissolve into microplastics in the machines and enter the water). Use paper cups and straws and recycle what you can. Compost garbage if possible (easy on a farm, hard in a condo). Use unbleached paper products (e.g., paper towels and napkins) to reduce water pollution from the whitening bleach. Small steps but still moving forward.

Toward Sustainability

If our society is to survive, we must develop a sustainable system and think and act for the long term. This means changing our worldview from one of conquering nature to one of living with nature. Either way, nature will ultimately prevail, so we might as well join in now.

We must change our exploitative policies so that we do not destroy the very resources we need in the future. Reforesting, restoration, and preservation of natural systems; conservation of water; use of traditional farming methods where appropriate; making Western farming less wasteful; cleaning up pollution; and developing green energy are important steps. We must also rein in the corporations and provide accessible education.

All this will require strong government regulations and adequate enforcement, but we fear governments are not yet under sufficient pressure to act. There is talk about acting, and plans have been proposed, but little action has been taken. Still, one cannot act without a plan, so plans are important first steps.

In 2015, the UN developed the sustainable development goals (SDGs), a collection of 17 combined objectives designed to realize a better and more sustainable future. The main goals are reducing poverty and hunger, the promotion of health, education, equality, clean water and good sanitation, clean energy, jobs and economic growth, development of industry, innovation and infrastructure, sustainability, responsible use of resources, climate action, respect for life, peace, justice, strong institutions, and partnerships. Many specific subgoals were also identified.

The SDG's were originally proposed and then extended to the world by two Colombian women working for the Colombian foreign ministry, Paula Caballero and Patti Londoño. They tell their story in their book *Redefining Development.*[27] They had to deal with attacks by people who thought such proposals were hopelessly impractical and visionary, and even "blasphemous" for challenging the conventional wisdom that throughput is the true good. Today, there seems no hope of achieving the SDGs by 2030, but smaller incremental steps may be possible by 2040 or 2050.[28]

Remembering and Practicing Morality

Traditional Systems Learned

Robin Wall Kimmerer is a biology professor and Native American (Potawatomi), living in upstate New York. Her bestselling book *Braiding Sweetgrass*[29] discusses at length the traditional use and management techniques of the Anishinabe (the large language group including the Potawatomi). It is all identical, so far as she tells it, to the traditions ENA encountered on the Northwest Coast and more or less what another Native American Biologist, Raymond Pierotti, describes from his background and experience.[30] Kimmerer relates it to such wider teachings as the rules she heard from depression-era elders, apparently both Indigenous and settler: "Use it up, wear it out, make it do, or do without."[31]

This general belief system is very widespread not only in the Americas but also in Siberia and neighboring parts of northern Eurasia. Kimmerer describes it as the Honorable Harvest:

> Know the ways of the ones who take care of you, so that you may take care of them.
> Introduce yourself. Be accountable, the one who comes asking for life.
> Ask permission before taking. Abide by the answer.
> Never take the first. Never take the last.
> Take only what you need.
> Take only that which is given.
> Never take more than half. Leave some for others.
> Harvest in a way that minimizes harm.
> Use it respectfully. Never waste what you have taken.
> Share.
> Give thanks for what you have been given.
> Give a gift, in reciprocity for what you have taken.
> Sustain the ones who sustain you and the earth will last forever.[32]

Specifically asking a plant for permission to harvest part of it and then thanking it and sometimes even leaving a small gift is a counsel of perfection that seems unknown outside of Canada and the neighboring northern United States, but the general idea of respect for the plants, animals, and minerals taken is a general rule. A plant may be giving itself to the harvester only in the latter's imagination, but the idea that game animals and fish offer themselves is less irrational; game will sometimes approach the hunter (probably out of curiosity), and fish do willingly take the baited hook. Assuming generosity on their part is probably wrong but certainly respectful.

As Kimmerer and many others have pointed out, respect is the key to this system. Respect is shown to animals by not overhunting them. Taking more than one needs, and taking more than the animal population can sustain, is the height of disrespect. Animal carcasses once taken are treated with respect also, which includes using everything (not discarding most of it as Euro-Americans do) or restoring any unusable parts to the wild. This is often clearly common sense, as when Northwest Coast people throw bones of salmon back into the river; the bones are needed by the young salmon for mineral nutrition. The reason that Pacific salmon die after spawning is that their bodies are necessary for this purpose, a point recognized by Native Americans there but extended to see the adult salmon as offering their lives to their descendants in an ultimate charity.

Less common-sensical is the humanization of plants and even rocks. Ordinary Westerners are familiar with such humanization of their pets, and even their cars, computers, and houses, but do not take it so seriously. The difference is that personalizing natural entities leads to treating them with care. You do not waste, slight, or casually destroy or throw away a real person.

This ideology of respect for the natural world, involving personalization of natural kinds, extends to Siberia and Mongolia. A young Mongolian lady from a traditional background told ENA (in the Gobi wilds) that she would like to collect pretty rocks but did not because "it would be disrespectful to the rocks to move them around for no good reason." Particularly good descriptions of this worldview exist, written by or recorded from local persons.[33]

A variant idea is that people and some animals are reincarnated in each other's form. The Haida of British Columbia and Alaska see men (male humans) as reincarnating in killer whales, and killer whales are duly reincarnated as men; women become other species. At the other end of the Americas, the Yshiro of Paraguay hold that peccaries (their main game animal) and humans become each other, so when a hunter bags a peccary, he is bagging a former human who is probably sacrificing himself or herself to feed his or her human relatives.[34]

There is plenty of evidence from oral tradition that this caretaking attitude was hard-won from experience. Kimmerer records tales of early-day people who wasted resources and, naturally, starved. Anderson and Pierotti made a thorough search of the Northwest Coast literature and found dozens of such stories. They fall into a common pattern, showing they have traded elements widely, but the basic story is simple: a village got carried away with the abundance of fish, plant food, game, or any other essential, overused it, even wasting and throwing away some—and then came a bad year and everyone starved, except the one or two young people who had been more sensible and treated the resources with due respect. The moral—do not take more than you need and waste nothing—is still drummed into the mind of every young Indigenous child living under even remotely traditional conditions, and to many even in the industrial cities of the north and west.

This, of course, refutes the extreme "noble savage" view that Indigenous people had little or no effect, or only good effect, on the environment. Every group, even remote ones with low population densities, has these stories of overharvest in the past and sober learning afterward.

Archaeology has dramatically confirmed the general message of such stories. In Polynesia and other Pacific islands, Patrick Kirch was the first to find evidence of settlement, followed by rapid overexploitation (often exterminating vulnerable species), followed by human population collapse, followed by slow rebuilding as people learned how to manage the system sustainably. Eventually the population is sustained at a higher level than earlier overexploitation had done, but only by introducing a wide range of useful plants and animals.[35] Ken Lertzman and others have made similar findings in the island groups.

Similarly, the complex Ancestral Puebloan societies in the Four Corners area of the American Southwest collapsed some 800 years ago,[36] apparently due to droughts. This event seems to have led to more sober, intensive, carefully planned agriculture later; at least the existing Pueblo societies of New Mexico and Arizona manage resources carefully.

If humans did indeed have a hand in exterminating much of the Pleistocene megafauna, we can assume a similar story. Many or most of the megafauna survived in Africa and east and southeast Asia, possibly because they and humans had more time to adapt to each other. Rapid extinctions took place in Australia, probably because of burning, and in the Americas, but many millennia after humans arrived, with a closer correlation with climate change than with human arrival. In any case, the humans adapted, finding ways to make a living from smaller game and more plant foods. Learning to process tannin-rich acorns and even poisonous cycads and buckeyes was a major achievement.

The incorporation of ecological knowledge and environmental morality in a single common discourse is widespread in traditional societies. In fact, its conspicuous lack in modern industrial society dates only from the "Scientific Revolution" of the 17th century, though the link was weakening in ancient times. Plantations and the attendant use of enslaved labor are the real source; when morality toward humans has been abandoned in the production process, morality toward the environment will follow. By contrast, the areas of the world not centered around huge-scale plantation production, even if they had slavery on a small scale, continued to respect and revere the natural world. Medieval literature personalizes plants, animals, and natural processes. The early Irish myths and texts repeat environmental morals strikingly similar to those in *Braiding Sweetgrass*.[37]

Southeast Asia's environmental state societies have been noted. Very widespread therein is a belief in the Rice Maiden, a beautiful young woman who is the spirit and life force of the region's staple food. She can be seen on misty nights, a white figure hovering over the fields, if you have enough imagination. The practical application is a widespread belief that she hides herself in the finest, fullest, most beautiful clusters of rice, which then must be saved for seed to plant next year's crop.[38] The obvious benefits of this for the future of rice production are perfectly clear to any modern plant breeder. Without personalized belief, the growers would be strongly tempted to share the finest rice heads in feasts. They presumably learned over centuries to save the best for seed, but not knowing modern genetics, interpreted this as a personal blessing of the rice spirits.

It is almost needless to say that here too, respect is a major concept. In the world of the Akha of south China and northern Southeast Asia, described in great detail by Wang Jianhua, it is taqheeq-e.[39] For the Akha, the world of humans, including gardens and fields, and the world of forests and forest spirits must remain in complementarity and must take care of each other. Respect includes, as usual, not overhunting or overexploiting wild foods, leaving forested areas, and growing crops with proper attention and care. Similar beliefs occur among the Karen,[40] Tibetans of forested areas,[41] and other groups.

Regard for sacred trees, groves, field spirits, and other religious agricultural botany led anthropologist Leslie Sponsel to innovate a whole field of "spiritual ecology," now flourishing worldwide, stimulating research on the sound ecological wisdom preserved in "religious" activities.[42] The Western world is far too prone to separate "religion" and "economy" as different fields of research. This led to massive failure to understand, appreciate, or use traditional wisdom; it was too often dismissed as "mere superstition," despite anthropological attention to the value of spiritual ecology ever since the mid-19th century.

A still different program exists in Aboriginal Australia, but it is broadly similar: environmental and ecological knowledge are embedded in spiritual belief and mythic texts and art. Respect and care for the environment are taught as part of religious initiation, though common sense extends well beyond this into ordinary everyday dealings. In the Dreamtime, the time before time that is also, somehow, ongoing now, non-human animals, plants, and even rocks may have a humanoid or at least intelligent and well-governed presence.[43] As Deborah Rose records, "Dingo makes us human."[44]

Landscapes were created by the activities of the Dreamtime beings. ENA has stood in a grove in northern Australia that wandered around the landscape looking for a home until it settled at this particular spot. Widely in this part of the world there are four genders: masculine, feminine, neuter, and useful plants. (The few useless plants are in the neuter gender.) This attributes a nonhuman and essentially different, but very real, personhood to the plants.[45] Relations with the environment are complex and intense, guided by personal relationships that were very different in the Dreamtime but are still relevant. The result is often incredibly skilled, detailed, and successful management of the landscape. Time after time, the government has had to resort to Aboriginal knowledge to save ecosystems and endangered species that were being ruined by Anglo-Australian management.[46]

Recently, traditional Aboriginal art portraying the Dreamtime and the environment has modernized, adopting acrylic paints, modern canvas and board, and freer interpretation of themes. This has produced some of the most compelling art in the world, as universally agreed by art critics from Australia to New York. Many good books on this art are out there.[47]

Africa's environmental relations are covered (slightly) in Chapter 7, but suffice it to say that similar spiritual ecologies are as widespread there as elsewhere. It would take many volumes to begin to cover this topic. Studies by Roderick McIntosh disclose a particularly complex and sophisticated ideology in central west Africa.[48] Ngozi Unuighe has made the comparison, all too similar to that everywhere else, between environmentally and ecologically sensitive local management and careless, unthinking, broad-brush approaches by outsiders, who, for instance, attempt to transplant industrial agriculture to Africa in spite of the extreme differences in soil and climate from those of Iowa or Germany.[49]

The major problems the world is facing can be solved not only in concept but also in local reality. Conserving water, biodiversity, forests, fish, and farmland are all done effectively around the world at local levels. This is a moral approach, and there is absolutely no excuse for governments continuing to support and subsidize destruction when they could do the same for sustainable management.

A small story opens some important history. Chris Madsen[50] reports that the wood duck is common again. When we were young, this beautiful duck was rare, and ENA was greatly excited to see his first one (rather anticlimactically at a park in downtown Los Angeles). No trouble now in finding them if you are near flooded forests or forest ponds. They were saved by hunters who were also conservationists—people who had the sense to realize that if there are no more ducks, you can't hunt ducks. Madsen traces the story to the early ethnographer and New York elite writer George Grinnell, who learned from the Native Americans how to respect wildlife and save it for the future.[51] Grinnell drew heavily on his Native American experiences and then on his New York connections, which included Teddy Roosevelt. Another hunter-conservationist-wildlife student, George Shiras III, was elected to the US House of Representatives and introduced the Migratory Bird Act in 1904 with Roosevelt's support. It became an international treaty in 1918, the first great landmark treaty in the field of biodiversity conservation. So, the key to wildlife conservation was ultimately Native American values being brought into public attention.

This was followed by an increasing number of important laws. All had basically the same story: an idea, often one acquired from a traditional and small-scale society, caught on with people who could raise the word in influential contexts. Eventually they started a popular movement that attracted the more responsible politicians, who could eventually persuade or browbeat the less responsible ones and get significant laws passed. No movement succeeds without laws, but they need real teeth and need to be proof against the evil machinations of corrupt and suborned courts.

The basics have been known for literally thousands of years. Traditional societies faced all the same problems and solved them well enough to endure. The solutions rarely work for a planet with several billion people, but they provide major ideas on which we can build. They show what can be done to simplify our production systems and live efficiently. Desert societies have rules for managing water, forest ones for maintaining trees, and coastal ones for maintaining fish. One exemplary researcher, Gary Nabhan, has been working for a lifetime on popularizing such ideas for postindustrial society, and his many books are storehouses of ideas.[52]

The guiding principle here is stated by African environmental scholar Ngozi Finette Unuighe: "The growth of capitalism spreading to every part of our planet was facilitated by the legal recognition of corporations as artificial persons capable of holding rights (codified in the U.S. by the Supreme Court 'Citizens United' decision), whereas life-giving species and components of the Earth, such as lakes, rivers, forests and mountains have been systematically denied their inherent rights to be and to flourish."[53] African traditions, like all other local and small-scale social traditions of management, privilege preserving the basic resources of life over profit by alien interlopers.

The obvious counter is to give natural things the same legal recognition as corporations. A pathbreaking essay in 1975, "Should Trees Have Standing?" by Christopher Stone,[54] opened a wide door to this. The pun in the title had a good deal to do with its initial popularity, but the content was seriously thought-provoking. Today, the Ganges River; a tree in Athens, Georgia; some cetaceans and a river in New Zealand; and several other natural kinds worldwide have legal status equivalent to the corporations. One hopes this will increase.

We know that investing in nature and natural benefits produces very high returns overall. It is not even economic to waste so much on throughput.

Richard Louv has long advocated getting children into nature and letting them experience and learn. He sees children today as usually growing up with "nature deficit disorder."[55] Adults too are healed by escaping the crowds and built environments. Many psychological studies confirm that open green and blue spaces soothe and ease us.[56] We were adapted for them, not for cities, and our nerves and souls do not forget.

An excellent overview by Natalie Seddon[57] lists major good and bad options. Good ones include bioswales, water catchment areas around large infrastructure projects, and now mandated in California. Also: "Indigenous land tilling and forest management…Farmer-led pasturelands management with a silvo-pastoral system…Participatory integrated watershed management…Community-based hybrid approach to build coastal resilience" with examples from South America, Africa, Naxi of China, and restoring reefs and mangroves everywhere. Bad is "Exclusion of Indigenous people and afforestation with non-native monoculture…. Controversial reforestation for carbon offsetting…. Top-down implementation of marine protected areas…. Peri-urban land grab for greenbelt development," with horrible examples from places like Mt. Elgon, Kenya, where locals were displaced for a reforestation that failed in ten years, and greenbelt land grabs around Medellin, Colombia.

Even small bits of natural environment are worth saving, though they are more apt to lose species than larger ones.[58] The larger an area, the higher the chance that animals will survive, but sometimes a tiny space is critical for animals that can successfully breed within such limits.

Reserves are necessary, but only to a point. Human pressure on many of them is enormous. Think of the crowds in the main canyons of Yosemite or Zion; no sensitive animals or birds will stay there. Many of the world's reserves, including most of the US public lands, allow logging and grazing. Not only reserves, but even carbon-mitigation plans and other ecologically necessary measures, must be carefully managed so as not to impact local communities; the effects of badly planned projects can be literally genocidal to small groups in fragile ecological situations, though the effects of continued industrial throughput are far worse.[59]

The great human ecologist Arturo Escobar has suggested many ways of improving the world by returning control to local people and by reconsidering our entire worldview. He advocates a relational ontology—a worldview based on looking at relationships, not only people-to-nature but also people-to-people—in the light of current design theories.[60] However, this approach requires that traditional peoples and societies also be preserved, a daunting task in the face of colonizing Western industrial society (recall the "Borg" from Chapter 4). If those local societies are destroyed, their knowledge of resource and landscape management, morality and ethics, philosophy of nature, local solutions, and traditions of care will also be lost, much to the detriment of us all.

From Suggestions to General Morality

Cooperation is a worldwide moral virtue—so obviously needed that it is respected everywhere.[61] It must be invoked first and last for any success in saving the planet.

In the new society, devoutly to be wished and coming inevitably, there will be new measures of prosperity. No longer will we look at "GDP," i.e., amount of goods and services produced and sold, but at overall GWP, i.e., overall prosperity as measured by health, sustainable use, equity, and happiness, as discussed earlier in this work. High status will not come from having more stuff or controlling it; status will come from the ability to help others, especially by advancing science and arts of life.

Morality will see all beings as worthy of respect. Animals will be a "community of subjects"[62] in the Kantian sense of beings to be taken as the ends of action, not as mere objects nor as mere means to an end.

Instead of hate and fear of nature, people will move toward the traditional small-scale social attitude of love and respect for nature and natural things, and for as much love and respect for people as we can manage. Anger, frustration, and resentment about not having *more* will be replaced by helpfulness, generosity, and simple satisfaction from socializing and beauty in life. People will no longer find their fulfillment in buying more junk.

Since Edward Barbier's book *A Global Green New Deal*,[63] the idea of a "green new deal" has been alive and flourishing. Some progress has been made in that direction. Barbier pointed out, in particular, the various economic and political-economic measures that hurt the environment and could be turned to advantage in saving it: subsidies, accounting, deficit spending, government economic and production policies, and so on.

Voluntary leaders—people doing environmental work and advocacy for their own satisfaction, not gain—made local resource management successful under difficult circumstances in Ecuador and Uganda. It was especially important in "creeping" problems—the "it won't hurt to take just one" cases—and when communication is sparse. Unselfish leadership helps develop trust and communication.

Exceptional leaders are needed,[64] but so is culture and society in general. Conformity is dangerous since most of us are all too wedded to ecologically bad habits: driving alone, eating high on the food chain, air-conditioning a big and poorly insulated house, not recycling.

Elinor Ostrom's key findings on management of common-pool resources remain golden. Full property rights to Indigenous peoples that are protecting resources are key in areas where they still have some control.[65] Indigenous rights and concern for endangered species is now being integrated, or "braided," into policies in Canada and elsewhere, as seen in Robin Wall Kimmerer's classic book *Braiding Sweetgrass*.[66] Ludomir Lozny and Thomas McGovern[67] recently brought together a large set of studies, many by former skeptics, that show how and when the management of the commons can actually work. They found that Ostrom was right: cooperation, limits (including secure tenure), and knowledge of the results of overdraft are usually enough. Hobbesian views of traditional humanity do not stand even a brief inspection. International treaties, cooperation, and agreements are critical, with methods for achieving these and the relative success of various treaties varying widely.[68]

Maximiliano Cox and Rodrigo Asún found that working with values, beliefs, and norms as one package worked to persuade people in Chile to save biodiversity. Values included biospheric, egoistic, and altruistic ones; beliefs were about what is right, good, and possible; norms were senses of personal obligation. Behaviors followed from invoking all these—including the egoistic ones—showing people that they would survive and thrive if biodiversity was spared. It was a neat model, and it worked.[69]

The ideology of our time, and indeed of Western society for the last few centuries, has been one of dominating nature or destroying natural landscapes and resources outright. This is no longer possible. Fortunately, almost all other societies have managed to find more ability to respect nonhuman landscapes, entities, and forces.[70] They have lived with varying degrees of respect and concern for the non-human world. At best, they respect all beings and treat them with consideration—not avoiding all use but considering wider values than immediate material returns.

Hope for the Future

Not all is ill with the modern world system. Industrial society is running its course and will either adapt or collapse in time as the Four Corners did in the 13th century. However, the modern world adopted much of its technology from an earlier time, when designing with nature and drawing on efficiency and natural gifts were still required by lack of an alternative. The most conspicuous successes in modern times have been in medicine; one can always point to the huge increase in life expectancy and decline in maternal and child deaths in the last century and a half.

Though this depended on modern industry in some ways—cold transport, better tools and equipment, rapid travel, and the like—it has been due most of all to two ancient ideas: inoculation and herbal medicine. Inoculation with smallpox was developed by the Chinese at some early (but uncertain) point, adopted in the west in the 18th century, and perfected by Edward Jenner and followers at that time. This idea of stimulating the body's natural immunity by exposing it to weakened strains of the natural pathogens has now saved hundreds of millions of lives. Smallpox has been totally eliminated, and polio and several other diseases could be eliminated in the near future if the anti-natural folly of "anti-vaxx" can be stopped.

Herbal medicine is older than humanity. Monkeys and apes use it, treating themselves internally and externally with antibiotic herbs. All human societies know healing plants, and some have mineral and animal cures too. Drug development in modern labs grew from such origins and still uses plants. Many modern cancer cures stem from chemicals found in *Catharanthus roseus* and yew foliage. Descendants of digitalin in foxglove—an English folk remedy—are still used for heart problems. Aspirin was originally derived from Spiraea (hence the name); the base chemical for it, salicylic acid, also occurs in willows and other plants and was used for fever reduction in folk societies.

Surgery, wound treatment, and antibiotics were all well developed in all or almost all human societies. Natural antibiotics used widely in ancient times included rose oil, thyme oil, honey, and many other effective compounds.

Medical progress not stemming from tradition but using traditional ideas includes modern minimalist surgery; we are learning to use less and less invasive and tissue-damaging procedures. Computers and other contemporary devices help, but the idea—do as little harm as possible—is ancient, already recognized by Hippocrates. And perhaps most important of all, the age-old best medical advice in the world is still quoted:

Use three Physicians still: first Doctor Quiet, then Doctor Merryman, and Doctor Diet.

(Sir John Haringon's Elizabethan translation from a Latin translation of a 10th-century Arab original.[71])

Naturally, this is not the only place where modern industry uses not only techniques of the past but also the whole traditional idea of doing more with less, wasting as little as possible, and saving, reusing, or restoring when those can be done. From construction and agriculture to hi-tech manufacturing, the ideas of efficiency and sustainability continue. The problem is that industrial society is still far too committed to maximizing throughput. We are afflicted with planned obsolescence, lawns, industrial agriculture, fast fashion, and hundreds of other ways of gratuitously destroying resources for no good purpose.

The End?

I have told you what you must do … You must now decide to do it.
(paraphrased from Guinan speaking to Captain Picard
on "Star Trek, The Next Generation," Yesterday's
Enterprise, Season 3, Episode 15)

The primary threat to our planet is clearly climate change and its ensuing climate crisis, but there are other problems, not all related to climate change. Sustainability is a theme that runs through many of the problems. Let's briefly review where we find ourselves.

The Climate Crisis. Due primarily to the influx of massive amounts of CO_2 into the atmosphere from burning fossil fuels, the climate is becoming hotter, more chaotic, and less predictable. Storms are more frequent and more powerful. Drought plagues some areas while others flood. The ice is melting and sea levels are rising. Coastal cities around the world will flood, and some entire island countries will be inundated. The heat in some regions has already caused farming to fail in some areas, creating a climate migrant crisis.

Mass Extinction. We are witnessing the beginning of a mass extinction event caused by humans. Climate change is one of the issues, but the overexploitation of forests (which adds to climate change) is a primary factor. We are destroying the habitats of millions of species that have nowhere else to go. The loss of biodiversity will have an adverse impact on all life just as the loss of rivets weakens the structure of an airliner. What will life on Earth look like in the future?

A Hot and Polluted Atmosphere. As the lower atmosphere gets hotter from climate change, people will begin to die from the heat; we already know that hundreds of thousands of people do so each year, and that number is increasing. Some areas may become so hot that they will be uninhabitable. Farming in those areas will be impossible, and people will have to move while the plants and animals will have to either adapt or die.

Related issues include the pollution of the air and its impact on human health. Considerable ambient pollution originates from the same sources (industries and vehicles) that contribute to climate change. Other factors include indoor pollution from inadequate ventilation and the use of poor-quality fuels. Millions die from these problems each year.

Lack of Fresh Water. Fresh water is unevenly distributed across the planet, and where it is available, it is still limited. Fresh surface water is being overexploited by agriculture and cities, and when that resource reaches its limits, underground fossil water is pumped out. Although that can work for a time, even those sources are being overexploited. Thus, many places are literally running out of water with no real plan to compensate. In underdeveloped areas, billions of people lack access even to clean drinking water and resulting health problems are rampant.

Unsustainable Farming. Virtually all of the eight billion people on the planet are dependent on food grown on farms. Contemporary industrialized farming is highly productive and provides a growing portion of the needed food supply. However, this system is dependent on fossil fuels, chemical fertilizers, pesticides, herbicides, large tracks of land stripped of its biodiversity, and enormous amounts of scarce water. It is also highly polluting and a major contributor to GHG emissions.

This system is unsustainable in its current form. How much longer it can operate is unknown, but it might suddenly crash, resulting in an inadequate and unstable food supply. As one might expect, that will create a major problem. Other traditional systems are sustainable but are under pressure to be more productive; that is, change to industrialized farming.

Deforestation. Forests are places of great biodiversity and store massive amounts of carbon. Their destruction impacts the webs of life, releases stored carbon back into the air, and prevents carbon already in the air from being taken up and stored. Drought and disease further weaken forests, and the recent spate of fires has been devastating.

So far, humans have destroyed about half of the existing forests, much of that in the last 50 years. While people use and need wood for a variety of purposes, much of the destruction has been to convert the forested areas into short-term pastures and farm fields, leaving the former forested areas barren after a few years. Such destruction is unsustainable for the health of the planet.

Polluted and Overfished Oceans. We have succeeded in polluting the oceans to the point that our trash can be found everywhere, from massive floating garbage patches to oil spills to sewage to microplastics in the guts of sea life. Nowhere is clean anymore. Such pollution has had an impact on marine food webs and marine foods used by humans.

People depend on foods from the oceans such as fish and shellfish. Hundreds of years of unregulated or poorly managed fisheries have resulted in so much overexploitation that some fisheries have collapsed while others may do so soon. We are on the brink of losing a major food source, and that could impact billions of people who depend on the sea.

Warfare. While some level of warfare is probably as old as modern humans, it was only recently that such conflict has had massive impacts on the environment. Mass destruction and casualties devastate regions, disrupting or destroying ecosystems and landscapes. Some of this damage is reparable, but there are also long-lasting effects. We now have the technological ability to destroy the entire planet.

A Sustainable Earth? Sustainable means the ability of a system to properly function over the long term. While this is a goal, the fact is that we do little that is sustainable. We use resources at rates that will exhaust them, with examples of fresh water, soils, and land for farming, forests, and fisheries being highlighted here. If we do not come to grips with our overexploitation of resources, we will one day suddenly find ourselves with little left and nowhere to go. Soon.

Solutions? We have endeavored to provide some ideas about how to deal with the existential problems that face us. Solutions include the use of old and new technologies, old and new farming methods, education, and social change. However, none of this can be accomplished without political will, meaning political change. Few of the current politicians seem to care, so the election of public officials that do care about the planet, its people, and environment is required.

This, and all the ecological changes needed, can come only in democracies. Autocratic regimes are invariably corrupted by giant corporate vested interests. There are a few benevolent despotisms out there, but very few and very small. The general rule is that the more autocratic a society becomes, the less it honors science, nature, or human lives. This can be seen in US politics today. The 2024 election of Claudia Sheinbaum, a Nobel prize-winning climate scientist, as the president of Mexico is a hopeful sign!

Fight for the planet!

Notes

1 Van der Linden, Sander. 2023. *Foolproof: Why Misinformation Infects Our Minds and How to Build Immunity*. New York: W. W. Norton.
2 Cook-Patton, Susan C., et al. 2020. "Mapping Carbon Accumulation Potential from Global Natural Forest Regrowth." *Nature* 585:545–550.
3 Mills, Maria B., et al. 2023. "Tropical Forests Post-Logging Are a Persistent Net Carbon Source to the Atmosphere." *Proceedings of the National Academy of Sciences* 120:e2214462120.
4 Hawken, Paul (ed.). 2017. *Drawdown: The Most Comprehensive Plan Ever Proposed to Reverse Global Warming*. New York: Penguin.
5 Jankowska, Emilia, et al. 2022. "Climate Benefits from Establishing Marine Protected Areas Targeted at Blue Carbon Solutions." *Proceedings of the National Academy of Sciences* 119:e2121705119.
6 Lewis, Simon L., et al. 2019. "Regenerate Natural Forests to Store Carbon." *Nature* 568(7750):25–28; Conniff, Richard. 2019. "The Last Resort." *Scientific American*, Jan., 52–59.
7 Jacobson, Mark. 2023. *No Miracles Needed: How Today's Technology Can Save Our Climate and Clean Our Air*. New edn. Cambridge: Cambridge University Press.
8 https://news.mit.edu/2023/carbon-dioxide-out-seawater-ocean-decorbonization-0216. Kim, Seoni Michael P. Nitzsche, Simon B. Rufer, Jack R. Lake, Kripa K. Varanasi, and T. Alan Hatton. 2023. "Asymmetric Chloride-Mediated Electrochemical Process for CO_2 Removal from Oceanwater." *Energy & Environmental Science* 16(5):2030–2044.
9 Francis, Pope. 2015. *Encyclical Letter Laudato Si*. Vatican: Papacy.

10 Al-Delaimy, Wael K., Veerabhadran Ramanathan, and Marcelo Sánchez Sorondo, eds. 2020. *Health of People, Health of Planet and Our Responsibility: Climate Change, Air Pollution and Health.* Cham, Switzerland: Springer.

11 https://www.iea.org/search/charts?q=fossil%20fuel%20demand.

12 Mitchell, Russ. "Rapid Decarbonization Is a Fantasy, Energy Historian Says." *Los Angeles Times,* Sept. 5, A9. See also Smil, Vaclav. 2017. *Energy and Civilization: A History* (2nd ed.). Cambridge: MIT Press.

13 https://news.engineering.pitt.edu/engineering-catalysts-that-turn-seawater-into-fuel/.

14 For the corruption of Equatorial Guinea, which has recently suffered a coup, see Appel, Hannah. 2019. *The Licit Life of Capitalism: US Oil in Equatorial Guinea.* Durham, NC: Duke University Press. For the other sources, Collier, Paul. 2010. *The Plundered Planet: Why We Must—and How We Can—Manage Nature for Global Prosperity.* Oxford: Oxford University Press; Damania, Ross, et al. 2023. *Nature's Frontiers: Achieving Sustainability, Prosperity and Efficiency with Natural Capital.* New York: World Bank Publications; Dore, Mohammed H. I., and Timothy D. Mount. 1999. *Global Environmental Economics: Equity and the Limits of Markets.* Oxford: Blackwell; Mazzucato, Mariana. 2018. *The Value of Everything: Making and Taking in the Global Economy.* London: Allen Lane; Wood, James W. 2020. *The Biodemography of Subsistence Farming.* Cambridge: Cambridge University Press.

15 Brook, Timothy. 2016. "Nine Sloughs: Profiling the Climate History of the Yuan and Ming Dynasties, 1260–1644." *Journal of Chinese History* 1:27–58.

16 Hoyer, Daniel, James S. Bennett, Jenny Reddish, Samantha Holder, Robert Howard, Majin Benam, Jill Levine, Francis Ludlow, Gary Feinman, and Peter Turchin. 2022. "Navigating Polycrisis: Long-Run Socio-Cultural Factors Shape Response to Changing Climate." *Philosophical Transactions B,* doi 10.1098/rstb.2022.0402; Anderson, E. N. 2019. *The East Asian World-System.* Cham, Switzerland: SpringerNature; Brook, Timothy. 2016. "Nine Sloughs: Profiling the Climate History of the Yuan and Ming Dynasties, 1260–1644." *Journal of Chinese History* 1:27–58.

17 Anderson, E. N. 2019. *The East Asian World-System.* Cham, Switzerland: SpringerNature; Brook, Timothy. 2016. "Nine Sloughs: Profiling the Climate History of the Yuan and Ming Dynasties, 1260–1644." *Journal of Chinese History* 1:27–58; Parker, Geoffrey. 2013. *Global Crisis: War, Climate Change and Catastrophe in the Seventeenth Century.* New Haven, CT: Yale University Press; Pei, Qing. 2021. *Climate Change Economics between Europe and China: Long-Term Economic Development of Divergence and Convergence.* Cham, Switzerland: SpringerNature.

18 Blake, Jonathan S., and Nils Gilman. 2024. *Children of a Modest Star: Planetary Thinking for an Age of Crisis.* Stanford, CA: Stanford University Press.

19 McKibben, Bill. 1989. *The End of Nature.* New York: Random House.

20 Lazarus, David. 2021. "Firms Can't Lie, So Why Can Politicians?" *Los Angeles Times,* Jan. 5, A6.

21 Wilkinson, Richard G., and Kate E. Pickett. 2024. "Why the World Cannot Afford the Rich." *Nature* 627:268–270.

22 On inequality and its dangers, see Piketty, Thomas. 2017. *Capital in the 21st Century.* Tr. Arthur Goldhammer. Cambridge, MA: Harvard University Press. On the whole current mess, see Turchin, Peter. 2023. *End Times: Elites, Counter-Elites, and the Path of Political Disintegration.* New York: Penguin Press.

23 Wilkinson, Richard, and Kate Pickett. 2009. *The Spirit Level: Why Greater Equality Makes Societies Stronger.* New York: Bloomsbury Press; Rawls, John. 1971. *A Theory of Justice.* Cambridge, MA: Harvard University Press; 2001. *Justice as Fairness: A Restatement.* Cambridge, MA: Harvard University Press.

24 Turchin, Peter. 2023. *End Times: Elites, Counter-Elites, and the Path of Political Disintegration.* New York: Penguin Press; Turchin, Peter, and Sergey Zefedov. Turchin, Peter. 2009. *Secular Cycles.* Princeton, NJ: Princeton University Press.

25 Ungar, Michael (ed.). 2021. *Multisystemic Resilience: Adaptation and Transformation in Contexts of Change.* New York: Oxford University Press.

26 Killean, Rachel, and Lauren Dempster. 2022. "Mass Violence, Environmental Harm, and the Limits of Transitional Justice." *Genocide Studies and Prevention* 16:11–39.

27 Caballero, Paula, and Patti Londoño. 2022. *Redefining Development: The Extraordinary Genesis of the Sustainable Development Goals.* New York: Lynne Rienner.

28 Fuso Nerini, Francesco, Mariana Mazzucato, Johan Rockström, Harro van Asselt, Jim W. Hall, Stelvia Matos, Åsa Persson, Benjamin Sovacool, Ricardo Vinuesa, and Jeffrey Sachs. 2024. "Extending the Sustainable Development Goals to 2050—A Road Map." *Nature* 630(8017):555–558.

29 Blake, Jonathan S., and Nils Gilman. 2024. *Children of a Modest Star: Planetary Thinking for an Age of Crisis.* Stanford, CA: Stanford University Press.

30 McKibben, Bill. 1989. *The End of Nature.* New York: Random House. 194.

31 Lazarus, David. 2021. "Firms Can't Lie, So Why Can Politicians?" *Los Angeles Times*, Jan. 5, A6.

32 Wilkinson, Richard G., and Kate E. Pickett. 2024. "Why the World Cannot Afford the Rich." *Nature* 627:268–270.

33 On inequality and its dangers, see Piketty, Thomas. 2017. *Capital in the 21st Century. Tr. Arthur Goldhammer.* Cambridge, MA: Harvard University Press. On the whole current mess, see Turchin, Peter. 2023. *End Times: Elites, Counter-Elites, and the Path of Political Disintegration.* New York: Penguin Press.

34 Wilkinson, Richard, and Kate Pickett. 2009. *The Spirit Level: Why Greater Equality Makes Societies Stronger.* New York: Bloomsbury Press; Rawls, John. 1971. *A Theory of Justice.* Cambridge, MA: Harvard University Press; Rawls, John. 2001. *Justice as Fairness: A Restatement.* Cambridge, MA: Harvard University Press.

35 Turchin, Peter. 2023. *End Times: Elites, Counter-Elites, and the Path of Political Disintegration.* New York: Penguin Press; Turchin, Peter, and Sergey Zefedov. 2009. *Secular Cycles.* Princeton, NJ: Princeton University Press.

36 Ungar, Michael (ed.). 2021. *Multisystemic Resilience: Adaptation and Transformation in Contexts of Change.* New York: Oxford University Press.

37 Killean, Rachel, and Lauren Dempster. 2022. "Mass Violence, Environmental Harm, and the Limits of Transitional Justice." *Genocide Studies and Prevention* 16:11–39.

38 Caballero, Paula, and Patti Londoño. 2022. *Redefining Development: The Extraordinary Genesis of the Sustainable Development Goals.* New York: Lynne Rienner.

39 Madson, Chris. 2015. "Comeback of the Wood Duck." *Ducks Unlimited*, Nov.-Dec., 74–81.

40 Merchant, Carolyn. 2016. *Spare the Birds! George Bird Grinnell and the First Audubon Society.* New Haven, CT: Yale University Press.

41 Nabhan, Gary Paul. 2013. *Growing Food in a Hotter, Drier Land: Lessons from Desert Farmers on Adapting to Climate Uncertainty.* White River Junction, VT: Chelsea Green. Nabhan, Gary Paul. 2014. *Ethnobiology for the Future: Linking Cultural and Ecological Diversity.* Tucson: University of Arizona Press; See also Nelson, Melissa K. and Dan Shilling (eds.) 2018. *Traditional Ecological Knowledge: Learning from Indigenous Practices for Environmental Sustainability.* Cambridge: Cambridge University Press; Peters, Charles M. 2018. *Managing the Wild: Stories of People and Plants and Tropical Forests.* New Haven, CT: Yale University Press; Watson, Julia. 2020. *Lo-TEK: Design by Radical Indigenism.* Cambridge, MA: Harvard University Press.

42 Unuighe, Ngozi Finette. 2020. "African Eco-Philosophy and Its Implications for Ecological Integrity in Africa." In *Ecological Integrity in Science and Law*, Laura Westra, Klaus Bosselmann, and Matteo Fermeglia (eds.), pp. 99–109. Cham, Switzerland: SpringerNature. See p. 100.

43 Stone, Christopher D. 1975. *Should Trees Have Standing?* New York: Avon.

44 Louv, Richard. 2005. *Last Child in the Woods: Saving Children from Nature-Deficit Disorder.* Chapel Hill: Algonquin Books of Chapel Hill; 2019. *Our Wild Calling: How*

Connecting with Animals Can Transform Our Lives—and Save Theirs. Chapel Hill, NC: Algonquin Books.

45 Johnson, Justin Andrew, et al. 2023. "Investing in Nature Can Improve Equity and Economic Returns." *Proceedings of the National Academy of Sciences* 120:2220401120; Weir, Kirsten. 2020. "Nurtured by Nature." *Monitor on Psychology*, April/May, 50–56; White, Matthew P., et al. 2019. "Spending at Least 120 Minutes a Week in Nature is Associated with Good Health and Wellbeing." *Scientific Reports* 9, article 7730.

46 Seddon, Nathalie. 2022. "Harnessing the Potential of Nature-Based Solutions for Mitigating and Adapting to Climate Change." *Science* 376:1410–1415.

47 Wintle, Brendan, Heini Kujala, Amy Whitehead, Alison Cameron, Sam Veloz, Aja Kukkala, Atte Molanen, Ascelin Gordon, Pia E. Lentini, Natasha C. R. Cadenhead, and Sarah A. Bekessy. 2019. "Global Synthesis of Conservation Studies Reveals the Importance of Small Habitat Patches for Biodiversity." *Proceedings of the National Academy of Sciences* 116:909–914.

48 Jones, Kendall R., Oscar Venter, Richard A. Fuller, James R. Allan, Sean L. Maxwell, Pablo Jose Negret, and James E. M. Watson. 2018. "One-third of Global Protected Land Is under Intense Human Pressure." *Science* 360:788–791; Paulose, Regina Menachery. 2022. "Death by a Thousand Cuts? Green Tech, Traditional Knowledge, and Genocide." *Genocide Studies and Prevention* 16:40–59.

49 Escobar, Arturo. 2018. *Designs for the Pluriverse: Radical Interdependence, Autonomy, and the Making of Worlds*. Durham, NC: Duke University Press; 2008. *Territories of Difference: Place, Movements, Life, Redes*. Durham, NC: Duke University Press.

50 Madson, Chris. 2015. "Comeback of the Wood Duck." *Ducks Unlimited*, Nov.-Dec., 74–81.

51 Merchant, Carolyn. 2016. *Spare the Birds! George Bird Grinnell and the First Audubon Society*. New Haven, CT: Yale University Press.

52 Nabhan, Gary Paul. 2013. *Growing Food in a Hotter, Drier Land: Lessons from Desert Farmers on Adapting to Climate Uncertainty*. White River Junction, VT: Chelsea Green. Nabhan, Gary Paul. 2014. *Ethnobiology for the Future: Linking Cultural and Ecological Diversity*. Tucson: University of Arizona Press; see also Nelson, Melissa K. and Dan Shilling (eds.) 2018. *Traditional Ecological Knowledge: Learning from Indigenous Practices for Environmental Sustainability*. Cambridge: Cambridge University Press; Peters, Charles M. 2018. *Managing the Wild: Stories of People and Plants and Tropical Forests*. New Haven, CT: Yale University Press; Watson, Julia. 2020. *Lo-TEK: Design by Radical Indigenism*. Cambridge, MA: Harvard University Press.

53 Unuighe, Ngozi Finette. 2020. "African Eco-Philosophy and Its Implications for Ecological Integrity in Africa." In *Ecological Integrity in Science and Law*, Laura Westra, Klaus Bosselmann, and Matteo Fermeglia (eds.), pp. 99–109. Cham, Switzerland: SpringerNature. See p. 100.

54 Stone, Christopher D. 1975. *Should Trees Have Standing?* New York: Avon.

55 Louv, Richard. 2005. *Last Child in the Woods: Saving Children from Nature-Deficit Disorder*. Chapel Hill: Algonquin Books of Chapel Hill; 2019. *Our Wild Calling: How Connecting with Animals Can Transform Our Lives—and Save Theirs*. Chapel Hill, NC: Algonquin Books.

56 Johnson, Justin Andrew, et al. 2023. "Investing in Nature Can Improve Equity and Economic Returns." *Proceedings of the National Academy of Sciences* 120:2220401120. Weir, Kirsten. 2020. "Nurtured by Nature." *Monitor on Psychology*, April/May, 50–56; White, Matthew P., et al. 2019. "Spending at Least 120 Minutes a Week in Nature is Associated with Good Health and Wellbeing." *Scientific Reports* 9, article 7730.

57 Seddon, Nathalie. 2022. "Harnessing the Potential of Nature-Based Solutions for Mitigating and Adapting to Climate Change." *Science* 376:1410–1415.

58 Wintle, Brendan, Heini Kujala, Amy Whitehead, Alison Cameron, Sam Veloz, Aja Kukkala, Atte Molanen, Ascelin Gordon, Pia E. Lentini, Natasha C. R. Cadenhead, and Sarah A. Bekessy. 2019. "Global Synthesis of Conservation Studies Reveals the

Importance of Small Habitat Patches for Biodiversity." *Proceedings of the National Academy of Sciences* 116:909–914.

59 Jones, Kendall R., Oscar Venter, Richard A. Fuller, James R. Allan, Sean L. Maxwell, Pablo Jose Negret, and James E. M. Watson. 2018. "One-third of Global Protected Land Is under Intense Human Pressure." *Science* 360:788–791; Paulose, Regina Menachery. 2022. "Death by a Thousand Cuts? Green Tech, Traditional Knowledge, and Genocide." *Genocide Studies and Prevention* 16:40–59.

60 Escobar, Arturo. 2018. *Designs for the Pluriverse: Radical Interdependence, Autonomy, and the Making of Worlds*. Durham, NC: Duke University Press; 2008. *Territories of Difference: Place, Movements, Life, Redes*. Durham, NC: Duke University Press.

61 Curry, Oliver Scott, Daniel Austin Mullins, and Harvey Whitehouse. 2019. "Is It Good to Cooperate? Testing the Theory of Morality-as-Cooperation in 60 Societies." *Current Anthropology* 60:47–69.

62 Waldau, Paul, and Kimberley Patton (eds.). 2007. *A Communion of Subjects: Animals in Religion, Science, and Ethics*. New York: Columbia University Press.

63 Barbier, Edward B. 2010. *A Global Green New Deal*. Cambridge: Cambridge University Press.

64 Amel, Elise, Christie Manning, Britain Scott, and Susan Koger. 2017. "Beyond the Roots of Human Inaction: Fostering Collective Effort toward Ecosystem Conservation." *Science* 356:275–279; Andersson, Kristen P., Kimberlee Chang, and Adriana Molina-Garzón. 2020. "Voluntary Leadership and the Emergence of Institutions for Self-Government." *Proceedings of the National Academy of Sciences* 117:27292–27299.

65 Anderson, E. N., and Raymond Pierotti. 2022. *Respect and Responsibility in Pacific Coast Indigenous Nations: The World Raven Makes*. Cham, Switzerland: SpringerNature; Baragwanath, Kathryn, and Ella Bayi. 2020. "Collective Property Rights Reduce Deforestation in the Brazilian Amazon." *Proceedings of the National Academy of Sciences* 117:20495–20502.

66 Kimmerer, Robin Wall. 2015. *Braiding Sweetgrass: Indigenous Wisdom, Scientific Knowledge and the Teachings of Plants*. Minneapolis, MN: Milkweed Editions; Lamb, Clayton, et al. 2023. "Braiding Indigenous Rights and Endangered Species Law." *Science* 380:694–696.

67 Lozny, Ludomir R., and Thomas H. McGovern (eds.). 2020. *Global Perspectives on Long Term Community Resource Management*. Cham, Switzerland: Springer.

68 Barrett, Scott. 2003. *Environment and Statecraft: The Strategy of Environmental Treaty-Making*. New York: Oxford University Press; Barrett, Scott. 2016. "Cooperation vs. Voluntarism and Enforcement in Sustaining International Environmental Cooperation." *Proceedings of the National Academy of Sciences* 113:14515–14522.

69 Cox, Maximiliano, and Rodrigo Asún. 2023. "Applicability of the Value-Belief-Norm Model to the Protection of Native Biodiversity in a District of Santiago, Chile." *Human Ecology* Review 27:93–114.

70 Jenkins, Willis, Mary Evelyn Tucker, and John Grim (eds.). 2016. *Routledge Handbook of Religion and Ecology*. New York: Routledge; Johnson, Leslie Main, and Eugene S. Hunn (eds.). 2010. *Landscape Ethnoecology: Concepts of Biotic and Physical Space*. New York: Berghahn.

71 Regimen Sanitatis Salernitanum. 1920 (orig. ca. 1600). *The School of Salernum: Regimen Sanitatis Salernitanum*. Trans. by Sir John Harington. New York: Paul B. Hoeber.

GLOSSARY

anthropocene a newly proposed epoch, beginning about 1950, defined by the massive human changes to the environment

anthropogenic human-caused changes in some system such as climate

anthropology the study of humans (biology, culture, language, past and present, etc.)

biodiversity the number and dominance of species present in an ecosystem

biomass the quantity (mass or weight) of a particular species or of all living matter within a specified area

biosphere the global environment and its interacting ecosystems

browsers animals who primarily eat the foliage of bushes and trees

calorie a measure of energy within something

carbohydrates sugars and starches found in foods

carbon the element at the center of the climate crisis

carbon capture various ways, natural and synthetic, to capture carbon and remove it from the air

carbon cycle how carbon moves through plants and animals and where it ends up

carbon dioxide one of the major greenhouse gases in the atmosphere

carbon footprint the amount of carbon emissions an entity (e.g., an individual, company, city, nation) is responsible for

carnivore an animal whose primary food is meat

chinampa a small farm field built in a marsh or lake

CO_2 carbon dioxide, a major greenhouse gas

common-pool resources resources that are not owned or regulated

cultural relativism the principle that all societies are valid, have the right to exist, and should not be judged.

domestication the selective breeding to increase the size or the quantities and qualities of seeds, roots, hair, or milk and to decrease their independence and aggressiveness

ecology the study of the connections between living things and their environment

ecosystem a living community tied together in a system

ecozone a specified environment based on its biology such as a forest, grasslands, or meadow

empirical science knowledge system based on tangible data

environment the surroundings within which things (e.g., an organism) interacts

epoch a broad time period such as the Pleistocene or Holocene

ethnocentrism the belief that your group is better than others

EU European Union

evolution change in a thing or system

feng shui the Chinese art of proper placement of materials in a space, from landscapes to rooms, to insure harmony

food chain the flow of nutrients through a system; the sequence of which organism is eating which

food web a series of food chains linked together

garden a small agricultural plot, tilled by hand

GMO genetically modified organism

genocide the killing, or attempted killing, of an entire ethnic group, religious group, or nation

grazers animals that mainly eat grasses and other low-growing plants

habitat the geographical place an organism lives

herbivore an animal that eats only plants

Hobbesian view that humans on their own lack a moral compass unless there is regulation

Holocene the geological epoch beginning at the end of the Pleistocene some 12,000 years ago to the present time

hunters and gatherers societies that make their primary living from wild foods and materials

ice ages the times within the Pleistocene when the ice sheets covered much of the northern portions of the northern hemisphere

intensive agriculture large-scale agriculture often involving the use of animal and/or mechanical labor, equipment, and water diversion techniques, production of a large surplus

irrigation the diversion of water from its natural source onto agricultural fields

Little Ice Age a major climatic cooling event between 700 and 250 years ago

monoculture growing a single species of a domesticated plant together in a field

niche what an organism eats and how it reproduces

omnivore animals that will eat most foods, plants, and animals

panarchy smaller cycles of a system nested in larger cycles, which are nested in still larger ones

permaculture permanent, sustainable agriculture growing many species

Pleistocene the geological epoch that encompassed the ice ages, from about 2.6 million to 12,000 years ago

protein a cluster of amino acids, broken down by the body to reuse its constituent amino acids

resource something used by an organism such as air, water, food, and materials

resource management the management of specific resources to make them more abundant or available

silviculture agriculture highlighting tree crops

slash-and-burn a farming method involving the cutting and burning of the natural vegetation from a small plot to clear it for the planting of crops.

state a large-scale society with complex social and political structures

stress a condition that makes a system less efficient and may force change

subsistence a complex system of making a living that includes the technologies and organizations necessary to procure foods and materials

sustainability taking no more than that a system can produce over a long time

swidden a sustainable system involving the rotation of slash-and-burn fields

tragedy of the commons the unregulated exploitation of a common-pool resource by a few users resulting in the exhaustion of the resource

transhumance the seasonal movement of animals from pasture to pasture

trauma a physical or psychological injury

UN United Nations

vitamins organic compounds necessary for certain body functions

Western industrial state societies, including the US, Europe, China, Japan, India, and others

INDEX

Note: *Italic* page numbers refer to figures.